From Everlasting to Everlasting

American University Studies

Series VII
Theology

Vol. 66

PETER LANG
New York · San Francisco · Bern
Frankfurt am Main · Paris · London

David Ingersoll Naglee

From Everlasting to Everlasting

John Wesley on Eternity and Time
Vol. 2

PETER LANG
New York · San Francisco · Bern
Frankfurt am Main · Paris · London

Library of Congress Cataloging-in-Publication Data

Naglee, David Ingersoll
  From everlasting to everlasting : John Wesley on
eternity and time / David Ingersoll Naglee.
    p. cm. — (American university studies. Series VII,
Theology and religion ; v. 65-66)
    Includes bibliographical references.
    1. Wesley, John, 1703-1791. 2. Eternity-History of
doctrines—18th century. 3. Time—Religious aspects—
Christianity-History of doctrines—18th century. I. Title.
II. Series.
BX8495.W5N28   1991   236'.21—dc20   90-47316
ISBN 0-8204-1114-0 (v. 2)
      0-8204-1183-3 (set)   CIP
ISSN 0740-0446

BX
8495
.W5
N28
1991
V. 2

© Peter Lang Publishing, Inc., New York 1991

Printed in the United States of America.

TABLE OF CONTENTS

Preface

John Wesley, an eighteenth century priest in the Church of England, became famous as one of the chief founders of a religious movement called Methodism. He was its primary leader - its chief inspiration, its galvanizing organizer, its tireless evangelist, its thoughtful apologist, its ubiquitous pastor, its erudite teacher, its prolific author, its moral and spiritual conscience, its social idealist, its finagling financier, its critical interpreter of Scripture and history, its model of duty, its visionary of destiny, its effervescent example of zeal and courage, and, hardly least of all, he was Methodism's embodiment of primitive Christianity. Yet, he never intended that his followers should leave the Church of England and begin a new church. Many did, however. The War of Independence in the colonies severed the American Methodists from Mother Church, and Wesley felt obligated to help them organize into a church after that conflict. In England, on the other hand, he kept a tight reign on his Methodists and kept them from separation. However, within several decades of his death, British Methodism also departed the Anglican communion to become a church on its own terms.

Presently, the Methodist bodies worldwide number well in excess of eleven million communicants. This contemporary phase of the movement, in the past few years, has shown a renewed interest in discovering the meaning of John Wesley - the man, the message, and the mission. The bicentennial of American Methodism in 1984 thrust that large denomination into the process. One important result has been the broadened understanding of Wesley's religious thought. Traditionally, Wesley's theology was treated in the narrow confines of soteriology, with emphasis on such themes as justification, regeneration, sanctification, and assurance. Seldom did scholarly works seriously treat subjects as eternity, creation, Chris-

tian living in terms of pilgrimage, stewardship, marriage, or eschatology.

In December 1980, I was in London finishing research on another Wesleyan manuscript. While at Evensong in Westminster
Abbey, as the choir sang a beautiful Advent anthem, a most un-
orthodox thought came to mind - "When you return home, read
Wesley's JOURNAL in reverse, from end to beginning." This idea
was preposterous at first. However, two months later, when the
reading was completed, a radically broadened understanding of
Wesley and his thought emerged. No great emphasis existed on
the famous "Aldersgate Experience." Traditional treatments always stressed that religious experience as being formative for
Wesley's thought, practice, and mission. Aldersgate, even for me,
colored everything Wesley said and did after 1738. But, in the re-
versed flow of his JOURNAL, it plays a very minimal part, covering less than a few years at most. If Aldersgate salvation, then, is
not the meaning of his theology, what is? This book seeks to answer that radical question, but the answer is not that radical. The
answer shows Wesley's commitment to a rather systematic theology
- not of his creation - but of his preference, inherited from late seventeenth century Anglicanism.

There are persons of infinite worth who have made this book
possible. Without their expertise and cooperation, my efforts
would have been quite minimal - Mr. Ian Pierson, Librarian of the
Society for the Propagation of the Gospel in Foreign Parts, London; Mrs. Charlene Baxter, Catalog Librarian at LaGrange College; and Stephen C. Hemphill, Director of Educational Seminars -
Cokesbury Bookstore, Atlanta, Georgia, who served as my reader
and literary adviser. To such belongs the kingdom of God!

David I. Naglee, Ph.D.,
Professor of Religion and Philosophy
LaGrange College
LaGrange, Georgia

"How advisable, by every possible means, to connect the ideas of time and eternity, so to associate them together, that the thought of one may never recur to your mind, without the thought of the other! It is our highest wisdom to associate the ideas of the visible and invisible world; to connect temporal and spiritual, mortal and immortal being."
- John Wesley, February 19, 1790

To Elfriede, My Wife

Beloved pilgrim with me,
Loved to all Eternity!

Chapter Seven

The Pilgrimage from Paradise Lost to Paradise Improved: The Pilgrim and Marriage

The pilgrim's journey through the time dimension is to be shared with persons of like precious faith. Wesley understood the Christian pilgrimage to possess this social feature. He repeatedly stated that true religion is never solitary religion. In 1789, two years before his death, he declared, "It is a blessed thing to have fellow-travellers to the New Jerusalem. If you cannot find any, you must make them; for none can travel that Road alone."[1] He employed his brother Charles's hymns upon this important theme -

> "Didst thou not make us one,
> That we might remain;
> Together travel on,
> And bear each other's pain;
> Till all thy utmost goodness prove,
> And rise renewed in perfect love.
>
> Surely thou didst unite
> Our kindred spirits here,
> That all hereafter might
> Meet at the marriage of the Lamb,
> And all thy gracious love proclaim."[2]

> "Together let us sweetly live,
> Together let us die;
> And each a starry crown receive,
> And reign above the sky."[3]

From Creation, God established marriage to be the ordinary re-lationship for adult pilgrims on their way into eternity *a parte post*. Ideally, marriage should prove sanctifying for both husband and wife, and it should bring a new generation of believers into the Christian way.

Unlike contemporary Methodism, John Wesley had a theology of marriage. It was based on biblical sources, and it was supported with seventeen centuries of Christian understanding and practice. A study of this theology of marriage, and its meaning for pilgrim-age, is vital to the discovery of a neglected facet of Wesley's thought and life. It goes deeper than popular treatments of his af-fairs with Miss Sophy Hopkey and Widow Vazeille.

BIBLICAL MARRIAGE

Marriage was an ordinance of creation, imposed on man and woman in their innocence, before sin entered their lives. It was es-tablished on the sixth day of creation and thus preceded the ordi-nance of the sabbath. Man was made first, then woman, being taken surgically from man's side.[4] Upon Adam's awakening, God presented the woman to him - she being "a piece of himself" - and they covenanted together to be companions. In token of Adam's acceptance of her, he gave her a name (later generations would give a ring as the token). Her name was *Isha*, or "woman" a "she-man, differing from man in sex only, not in nature; made of man, and joined with man."[5] Wesley explicitly affirmed that Isha (Eve) was "equal with man" by both creation and marriage. Marriage did not subjugate her to her husband.[6] It was the first sin that brought her into an inferior relationship with her husband. Then God said, "he shall rule over thee." Moreover, Eve was cursed with sorrow in child-bearing.[7] After being expelled from the Garden of Eden, Wesley continued, Adam and Eve had many sons and daughters.[8] Since they were the only humans on the face of the earth, they married one another. Lamech "took two wives" and thus perverted the true meaning of marriage since the "original law of marriage (declared) that two only should be one."[9] Indeed, when Adam and

Eve married, Adam explicitly pronounced, "Therefore shall a man leave his father and mother, and shall cleave unto his wife: and they shall be one flesh."[10]

Wesley argued further that monogamous marriage alone can enjoy God's blessing when it is owned publicly. Observe the case of Abram and Sarah when they kept their marriage secret from Abimelech. As a result, God did not grant them children at first. He would not bless them because they would not "own" their marriage - "Why should God own it?"[11] Marriage, therefore, must always maintain a public appearance. Wesley stated that Boaz and Ruth engaged in the first stage of Hebrew marriage (*Erusin*) by appearing before ten witnesses, and then afterwards they consummated the marriage agreement in the privacy of the marriage chamber (*nissuin*).[12] The legal contract made before witnesses remained public knowledge, and the marriage was judged by how well these legal vows were kept. Whatever transpired in privacy between husband and wife was not the business of the public.

Wesley understood Old Testament marriage to be a process rather than an event. The fathers of "houses" arranged marriages for their children. Laban arranged a marriage for his daughter, Rachel, with his nephew, Jacob. The arrangement was finalized before ten witnesses, with a *mohar* (price to be paid the bride's father for the bride) being agreed upon. Jacob agreed to work for Laban seven years, at the end of which he would get Rachel to take into the chamber for the consummation of the marriage. The first stage (*erusin*) was the legal side of marriage. And the second stage (*nissuin*) was the spiritual side. The first stage was public, and the second was private. The first stage featured vows, agreement on the mohar, and the token. At its conclusion the man and woman were legally married, but they did not live together. They were "espoused" and only death or divorce could separate them. Wesley illustrated the espousal side of marriage with the stories of Hosea,[13] and Joseph and Mary.[14] He also explained, "It was customary among the Jews for persons that married to contract before witnesses some time before."[15] The second stage - generally a month to a year after the first - featured the act of sexual intercourse, known in the Bible of Wesley as "knowing" one's wife. Wesley was much too modest to explain this stage of marriage. He usually referred to it as "consummating" the marriage - a customary expression used from the beginning of the Medieval period. Nevertheless, he understood that it is this second stage of marriage that makes "the two as one." It is in this act between husband and wife that God joins together more than bodies of flesh, but also hearts and souls and minds. Hence the husband indeed comes to "know" his wife, and she comes to "know" him.[16] But the *nissuin* act brings more than "mutual comfort." It serves "as well as for the preservation and increase of their kind." This is clear from the first marriage -

> "Adam and Eve were both made immediately by the hand of God, both made in God's likeness; and therefore between the sexes there is not that great difference and inequality which some imagine. . .God blessed them. It is usual for parents to bless their children, so God the common Father blessed his. . .chiefly to the blessing of increase, not excluding other blessings."[17]

There were in the Old Testament, Wesley recognized, but two ways of terminating a marriage, whether the marriage was in the first or second stage: death or divorce. Death is final for marriage, leaving the surviving spouse free to remarry. As proof of this, Wesley cited the remarriage of Abraham to Keturah after the death of Sarah (Genesis XXV, 1ff). The commentary of the *EXPLANATORY NOTES UPON THE OLD TESTAMENT*, on this passage, reads: "His (Abraham's) family wanted a governess and it was not good for him to be thus alone."[18] Death terminates marriages. Marriages do not extend into the afterlife. Nor are marriages made there.[19] Death ends marriages in both the Jewish and Christian dispensations.

Divorce in the Old Testament, as Wesley understood it, was a device invented by Moses to terminate marriages that were doomed to failure for one reason or another:

"When a man hath taken a wife, and married her, and it come to pass that she find no favour in his eyes, because he hath found some uncleanness in her; then let him write her a bill of divorcement, and give it in her hand, and send her out of his house. And when she is departed out of his house she may go and be another man's wife. And if the latter husband hate her, and write her a bill of divorcement, and giveth her it in her hand, and sendeth her out of his house: or if the latter husband die, which took her to be his wife: Her former husband, which sent her away, may not take her again to be his wife after that she is defiled: for that is abomination before the Lord." (Deuteronomy XXIV, 1-4).

Wesley's commentary on this important passage reads -

"*Some Uncleanness* - Some hateful thing, some distemper of the body or quality of mind not observed before marriage: or some light carriage, as this phrase commonly signifies, but not amounting to adultery. *Let him write* - This is not a command as some of the Jews understood it, nor an allowance and approbation, but merely a permission of that practice for prevention of greater mischiefs, and this only until *the time of reformation*, till the coming of the *Messiah* when things were to return to their first institution and purest condition. V. 4. *May not* - This is the punishment of his levity and injustice in putting her away without sufficient cause, which by this offer he now acknowledgeth. *Defiled* - Not absolutely, as if her second marriage were a sin, but with respect to her first husband, to whom she is as a defiled or unclean woman, that is, forbidden things; forbidden are accounted and called *unclean*, Judges XIII, 7, because they may no more be touched or used than an unclean thing."[20]

Excluded from Wesley's definition of this "uncleanness" was "adultery." The great commandment - "Thou shalt not commit adultery" - "forbids all acts of uncleanness, with all those desires which produce those acts and war against the soul."[21] In his commentary on the same commandment, but in Deuteronomy, Wesley added "this commandment requires one to prevent every kind of unchastity."[22] In the Old Testament, Wesley argued, adultery was the sexual violation of one's wife by another. As such, adultery was the breaking of the principle of marriage as the "two as one flesh." An erring spouse, whether husband or wife, destroys the basis of the legal contract as well as the spiritual bond of the marriage.

Wesley observed that wives of the espousal relationship, when proven to be adulteresses, were condemned to stoning by the law of Moses, whereas, wives of the consummated marriage were simply put to death in an unspecified manner.[23]

Wesley rejected Mosaic divorce, based upon the principle of uncleanness, as an undermining of marriage as it was originally. On the other hand, he recognized that God allows divorce based upon the infidelity of adultery - adultery having already severed the bond of oneness. This is best illustrated by Wesley's commentaries on the teachings of Jesus upon the subject. It will be remembered that Wesley held a view that the Messiah (Christ) would, at his coming, restore marriage to its "first institution and purest condition," rescuing it from the permissive allowance of Moses.[24] Christ clearly did this, Wesley argued. In the Sermon on the Mount (Matthew V, 31), Christ rejected the Mosaic practice of issuing bills of divorcement for "trifling" reasons, causing the wife "to commit adultery. . .If she marry again."[25] God does not accept the termination of a marriage upon the conditions of Deuteronomy XXIV. For this reason, Jesus taught, any remarriage after such a divorce is adulterous. In another passage (Matthew XIX, 3), Wesley expanded this interpretation. The passage reads -

"And the Pharisees came to him, tempting him, and saying, Is it lawful for a man to put away his wife for every cause? And he answering said to them, Have ye not read, that he who made them made them male and female from the beginning. And said, For this cause a man shall leave his father and mother, and cleave to his wife: and they twain shall be one flesh? Wherefore they are not more twain, but one flesh. What therefore God hath joined together, let not man put asunder. They say to him, Why then did Moses command to give a writing of divorce, and put her away? He saith to them, Because of the hardness of your hearts Moses permitted you to put away your wives: but from the beginning it was not so. And I say unto you, Whosoever shall put away his wife, except for whoredom (adultery), and marry another, committeth adultery: and he that marrieth her that is put away committeth adultery."[26]

Wesley's commentary on this passage observes that the Pharisees' question sprang from the argument of Moses concerning uncleanness. While they reduced Mosaic "uncleanness" to being synonymous with "for every cause," Wesley reduced it to the husband's employment of "anything he dislikes in her." But Jesus rejected such a foundation, arguing from the Mosaic story of the origin of marriage in the Garden of Eden -

"Our Lord confutes them by the very words of Moses. He who made them made them male and female from the beginning - At least from the beginning of the Mosaic creation. And where do we read of any other? Does it not follow that God's making Eve was a part of His original design, and not a consequence of Adam's beginning to fall? By making them one man and one woman, He condemned polygamy; by making them 'one flesh' He condemned divorce. . .'Why did Moses command?' - Christ replies, 'Moses permitted,' not 'commanded,' it, because 'of the hardness of your hearts'; because neither your fathers nor you could bear the more excellent way."[27]

Originally, marriage was designed for permanency. When adultery entered the human experience, God reluctantly allowed the marriage to be terminated in divorce. But the permissiveness of Deuteronomy XXIV expanded the practice of divorce beyond the point of divine toleration, and it broadened the justification of trifling issues. Wesley's sermon, *ON THE SERMON ON THE MOUNT - III*, provides an interesting summation of this aspect of divorce -

"It has been said, Whosoever will put away his wife, let him give her a writing of divorcement: And then all was well; though he alleged no cause, but that he did not like her, or liked another better. 'But I say unto you, that whosoever shall put away his wife, saving for the cause of fornication,' (that is, adultery; the word *porneia* signifying unchastity in general, either in the married or unmarried state) 'causeth her to commit adultery,' if she marry again: 'And whosoever shall marry her that is put away committeth adultery.'"[28]

It should be noted, at this juncture of Wesley's argument, that a divorce based on considerations other than "adultery" is, before God, no divorce at all. Hence the wife, sent out of a man's house, with a divorce paper in hand, is still his wife before God. Consequently, if she married another, she enters into both adultery and polygamous relationship. Moreover, while Moses in Deuteronomy XXIV allowed such a second marriage without sin, Christ forbids the remarriage of a woman put away for a cause other than adultery. The next paragraph of Wesley's sermon makes this very explicit -

> "All polygamy is clearly forbidden in these words, wherein our Lord expressly declares, that for any woman who has a husband alive, to marry again is adultery. By parity of reason, it is adultery for any man to marry again, so long as he has a wife alive, yea, although they were divorced; *unless* that divorce had been for the cause of adultery: *In that only case there is no scripture which forbids to marry again.*"[29]

Marriage, in the New Testament sense, is a more excellent way when compared to marriage under the Mosaic dispensation. The reason for this, according to Wesley, lies in the fact that Christ restored marriage to its initial condition, as it was before the Fall.[30] Christ, in Christian marriage, restores the wife to a place of equality with her husband. Subjection is not the abiding nature of marriage. The subjection of Eve to her husband was the consequence of sin. The Second Adam, however, has provided an atonement for all sin and thus makes husband and wife truly one, not two with one in arbitrary bondage to the other. Wesley viewed Christian marriage as existing in a process of grace by which the two begin their relationship with espousal and later sexual consummation, being blessed by God who mystically joins them together. While Christ restores equality to their marriage, equality is the end (*telos*) to be attained. It is not actually present at the beginning of marriage, but the couple may come to achieve it as they pursue it with love and spiritual earnestness. At the beginning of their marriage it is given to them as an inheritance to be claimed later when their marriage has come of age. Even marriage must "go on to perfection" - the perfection of final equality.

Wesley's commentary on Ephesians V, 23-33 is an eloquent affirmation of Christian marriage in teleological ascent to actual equality. (1) Vss. 22-24: "Wives submit yourselves to your own husbands, as unto the Lord." Wesley observed, "Unless where God forbids."[31] The husband does not stand as the supreme authority over the Christian wife. He has only a limited authority. God has placed certain limits upon the wife's subjection to her husband. There are some areas of a spouse's life that a husband has no right to invade. For instance, a wife during her menstrual period is not subject to her husband's sexual desires.[32] Wesley saw this bit of Mosaic legislation as not being abrogated by the Gospel dispensation.[33] In all other matters, however, the wife must recognize that "the will of the husband is a law to the wife." Wesley's point is, that in this subjection the will of the husband is "a law," not "the law." Christ's law or will is supreme, exhibited in relation to his wife (the Church) in sacrificial love and not in chauvinistic demands. The husband, being much less than Christ, must learn the loving ways of Christ in relation to his spouse. He must create, through sacrificial love, an atmosphere of mutuality in which the wife will find it not grievous to want to please her husband in all things. She will then be subject to her husband "as unto the Lord." As the Church is subject to her husband, Christ, so the Christian wife is subject to her husband, "he being the head of the wife, as Christ is the head of the Church."[34] Wesley's point must not be overlooked: The headship of the husband over the wife is one of love and not arbitrary power, because Christ's headship of his Church is one of love and nothing else. In a sermon on family religion Wesley was most emphatic on this point -

"I cannot find in the Bible that a husband has authority to strike his wife on any account, even suppose she struck him first, unless his life were in immediate danger. I never have known one instance yet of a wife that was mended thereby. I have heard, indeed, of some such instances; but as I did not see them, I do not believe them. It seems to me, all that can be done in this case is to be done partly by example, partly by argument or persuasion, each applied in such a manner as is dictated by Christian prudence."[35]

(2) Vss. 25-31: "Husbands, love your wives, even as Christ loved the Church, and gave up himself for it." Wesley referred to Christ's

love for the Church as "the true model of conjugal affection." The model must be emulated by every Christian husband. Each husband should be as Christ, giving up himself for his wife in loving care and concern. As Christ expended the best efforts for improving his wife, the Church (vss. 26-28), so should the Christian husband actively increase the well-being of his wife through nourishing and cherishing her in all things, as he does his own body. To Paul's words - "He that loveth his wife loveth himself" - Wesley added this comment: "Which is not a sin, but an indisputable duty."[36] The logic behind such an interpretation is familiar - In marriage the two are made one. The body of the wife is also the body of the husband, and his body is her body. Paul had said, "No one ever hated his own flesh, but nourisheth and cherisheth it." Hence, it is a duty to love the wife as one's self. "For this cause shall a man leave his father and mother, and shall be joined to his wife, and they two shall be one flesh" - a most "intimate union."[37] (3) Vss. 32-33: "This is a great mystery: I mean concerning Christ and the Church. But let every one of you in particular so love his wife as himself; and let the wife reverence her husband."[38] The essence of the husband's love (*agape*) is self-sacrificing for the well-being of his wife. The essence of the wife's subjection is "reverence" for her husband.

Wesley recognized other important facets of New Testament marriage. St. Paul's discourse on the subject, instructing the Corinthians (I Corinthians VII), was given a prominent treatment in Wesley's *EXPLANATORY NOTES UPON THE NEW TESTAMENT*. His main points are: (1) In addition to the many advantages of marriage, one surely escapes being a fornicator thereby. The sexual touching of a woman is permissible only in the bond of marriage. Therefore, "let every man have his own wife." Such an exhortation rules out polygamy and affirms monogamy as the Christian norm. It also rules out living with a woman outside of marriage - "Let not. . .persons fancy that there is any perfection in living with each other. . .unmarried."[39] Consequently, Paul argued, "Let every man have his own wife, and let every woman have her own husband." The husband possesses the wife and the wife the husband, her body under his power and his body under her power.[40] (2) Christian couples must not separate from one another, except by mutual agreement for a brief time, and that only for the practice of piety, after which they must come back together. Should they not become reunited, Satan will surely gain hold of them through sexual temptations.[41] (3) A successful Christian marriage depends upon a divine gift of grace - "But everyone hath his proper gift from God."[42] (4) In cases of unsuccessful marriage, the wife should not leave her husband, nor should he divorce his wife, unless the marriage has been violated by adultery.[43] (5) In marriages between believer and unbeliever, the believing husband must not divorce his unbelieving wife if she consents to live with him. Wesley added, "The Jews, indeed, were obliged of old to put away their idolatrous wives (Ezra X, 3); but their case was quite different. They were absolutely forbid to marry idolatrous women; but the persons here spoken of were married while they were both in a state of heathenism." Paul, speaking to gentiles who allowed wives to divorce husbands, added, "And the wife who hath an unbelieving husband, that consenteth to live with her, let her not put him away." The reason for this injunction, as Paul and Wesley saw it, was based on the divine gift given in marriage - by which the unbelieving husband is sanctified (converted) by the believing wife and vice-versa.[44] Then Wesley carried Paul's argument a step further - "Else your children would have been brought up heathen; whereas now they are Christians."[45] One believing spouse sanctifies the entire family. Moreover, wives are to "guide the family."[46]

ANGLICAN SERMONS ON MARRIAGE

Besides exegeting the Bible for a theology of marriage, Wesley studied the subject in the historical unfolding of the Church, through the Fathers, the Middle Ages, the Reformers, and the Church of England. In January 1753, he reprinted William Whateley's *DIRECTIONS FOR MARRIED PERSONS*, adding his own preface to the work, hoping that Methodist couples would benefit from reading the entire work.[47] Originally, Whateley's

work was entitled, *A BRIDEBUSH OR A WEDDING SERMON.*[48]
Wesley found in this work the scarlet thread of connubial reciproc-
ity - "There is a mutual bond of duty standing betwixt man and wife.
They are indebted each to the other in a reciprocal debt."[49] Thus
man and wife are bound to one another in chaste fidelity, keeping
"each ones body each for other."[50] The remainder of the sermon is
lengthy and exhausting of biblical content. It is more a book than
an actual sermon to be delivered at a marriage ceremony. If it had
been delivered at such an occasion, the ceremony would have
taken nearly three hours to complete. Nevertheless, Wesley approved of its content and desired that his followers read and master
its directions for successful marriage. Some of Whateley's phrase-
ology finds its way into Wesley's own writings on the subject, as do
some of his arguments and illustrations.

William Whateley wrote another major work on Christian marriage, a book which Wesley also used in his teaching - *A CARE
CLOTH OR A TREATISE OF THE CUMBERS AND TROUBLES
OF MARRIAGE.* This work found its rootage in I Corinthian VII,
39-40. In encyclopedic fashion, it traces the attitudes and actions
which allow a marriage to drift into the rocky surf of life - from un-
preparedness, to carelessness in duty, to impatience, to frivolity,
and so on.[51]

However, the greatest sermon on marriage in the Church of
England Tradition, and well-known to Wesley - finding its way into
his many discussions on the subject, was *A HOMILY OF THE
STATE OF MATRIMONY*, one of the standard sermons of the
Church.[52] Regarding the *HOMILIES* as a source of authority for
faith and practice,[53] Wesley recognized its biblical thoroughness -
marriage "is instituted of God, to the intent that man and woman
should live lawfully in a perpetual friendship to bring forth fruit,
and to avoid fornication." The "fruit" includes children - "that they
many be brought up by the parents godly, in the knowledge of
God's word. . .that finally many might enjoy that everlasting immortality. . .as to increase the kingdom of God."[54] The "perpetual
friendship" of marriage is constantly threatened by division and dis-
cord. Consequently, "married persons must apply their minds in

most earnest wise to concord, and must crave continually of God the help of his Holy Spirit, so to rule their hearts and to knit their minds together."[55] Prayer must be the practice or habit of husband and wife, "lest hate and debate do arise betwixt them." Without fervent prayer, they cannot detect the instigation "of the ghostly enemy," Satan. "Rough and sharp words" bring provocations that often lead to "stripes" and miseries and sorrows. Wife-beating is clearly the instigation of the devil.[56] Wesley made an explicit comment on this assertion -

"I cannot find in the Bible that a husband has authority to strike his wife on any account, even suppose she struck him first, unless his life were in immediate danger. I never have known one instance yet of a wife that was mended thereby...It seems to me, all that can be done in this case is to be done partly by example, partly by argument or persuasion, each applied in such a manner as is dictated by Christian prudence. If evil can be overcome, it must be overcome by good. It cannot be overcome by evil; We cannot beat the devil with his own weapons."[57]

The HOMILY turns from the subject of prayer to the duties of a Christian husband. Wesley's commentaries on I Corinthians VII, Ephesians V, and I Peter III, reflect this passage. "The husband. . . ought to be the leader and author of love, in cherishing and increasing concord; which then shall take place if he will use moderation, and not tyranny, and if he yield anything to the woman." The Gospel Spirit obliges the husband to be the "leader and author of love." It is the divine imperative binding all Christian husbands to the will of Christ who leads and loves his wife, the Church. Since the wife is a "weak creature," the husband "ought to wink at some things, and must gently expound all things, and to forbear."[58] Moreover, as St. Peter taught, husbands are required to employ reasoning and not fighting. "Yea, he saith more, that the woman ought to have a certain honour attributed to her."[59] The Christian man who brings these things into his marriage both pleases God and enriches his relationship with his wife.

The *HOMILY* also delineates the duties of the Christian wife as the appropriate response to a loving and considerate husband. The text for this response is I Peter III, 1 - "Ye wives, be ye in subjection to obey your own husbands." To "obey" the husband implies regarding his requests and perceiving what his needs are, thus bringing honour to God and peace to the home.[60] Yet, if a husband lacks gentleness, or a wife is troublesome, the other partner is obligated by God to remain loyal to his or her duties. The neglect of a husband's duties should not give the wife an excuse to neglect her duties, nor the husband his.[61] Never, never should a man "beat his wife; God forbid that - for that is the greatest shame that can be, not so much to her that is beaten, as to him that doeth the deed. . . (we) may well liken a man (if we may call him a man, rather than a wild beast) to a killer of his father or his mother. . .(is he not) a bedlam-man, who goeth about to overturn all that he hath at home?"[62] Such a violent act against one's wife is "extreme madness." Better for that husband that the ground would "open and swallow him."[63] In rebuttal to a protesting husband, the HOMILY argues - "But if thou shouldest beat her, thou shalt increase her evil affections: for frowardness and sharpness is not amended with frowardness, but with softness and gentleness."[64] An ounce of prevention lies in careful preparation for marriage. To marry in haste is to repent in leisure. "Before all things. . .a man (must) do his best. . .to get him a good wife."[65]

Wesley recognized the wisdom of careful selection for marriage. The HOMILY's emphasis was actually a New Testament one, and Wesley treated the subject in his sermon of *FRIENDSHIP WITH THE WORLD*. Above all relationships, he argued, the marriage contract is the most solemn, requiring great preparation. Entering into marriage without thorough preparation is "the most horrid folly, the most deplorable madness, that a child of God could possibly plunge into."[66] The Christian is solemnly warned to avoid marriage with an unbeliever - "The prohibition is so absolute and peremptory: 'Be not unequally yoked with an unbeliever.' Nothing can be more express."[67] Alas, Wesley argued, many enter into marriage with an unbeliever on the following line of reasoning - "I grant. . .the person I am about to marry is not a religious person. She does not make any pretensions to it. She has little thought about it. But she is a beautiful creature. She is extremely agree-able, and, I think, will make me a lovely companion." To which Wesley countered, "This is a snare, indeed! Perhaps one of the greatest that human nature is liable to."[68] In his sermon *ON A SINGLE EYE*, Wesley expanded this theme to include parental domination in marriage selection for their daughters - "seeking to marry them well." The passage is eloquent and deserves full ren-dering -

"How great is the darkness of that execrable wretch (I can give him no better title, be he rich or poor) who will sell his own child to the devil, who will barter her own eternal happiness for any quantity of gold or silver! What a monster would any man be accounted, who devoured the flesh of his own offspring! And is he not as great a monster who, by his own act and deed, gives her to be devoured by the roaring lion? As he certainly does (so far as is in his power) who marries her to any ungodly man. 'But he is rich; but he has ten thousand pounds!' What, if it were a hundred thousand? The more the worse; the less probability will she have of escaping the damnation of hell. With what face wilt thou look upon her, when she tells thee in the realms below, 'Thou hast plunged me into this place of torment. Hadst thou given me to a good man, however poor, I might have now been in Abraham's bo-som. But, O! what have riches profited me? They have sunk both me and thee into hell!'"[69]

He further emphasized this lesson in his famous sermon entitled, *ON FAMILY RELIGION*. This passage is no less eloquent and powerful, also deserving of citation -

"Your son or daughter is now of age to marry, and desires your advice rela-tive to it. Now you know what the world calls a good match, - one whereby much money is gained. Undoubtedly it is so, if it be true that money always brings happiness: But I doubt it is not true; money seldom brings happiness, either in this world or the world to come. Then let no man deceive you with vain words; riches and happiness seldom dwell together. Therefore, if you are wise, you will not seek riches for your children by their marriage. See that your eye be single in this also: Aim simply at the glory of God, and the real happiness of your children, both in time and eternity. It is a melancholy thing to see how Christian parents rejoice in selling their son or their daughter to a wealthy Heathen! And do you seriously call this a good match? Thou fool, by parity of reason, thou mayest call hell a good lodging, and the devil a good master. O learn a better lesson from a better Master! 'Seek ye first the kingdom of God and his righteousness,' both for thyself and thy children; 'and all other things shall be added unto you.'"[70]

THE SERVICE OF HOLY MATRIMONY

A priest of the Church of England, Wesley found THE FORM OF SOLEMNIZATION OF MATRIMONY both definitive and spiritual. He preferred *THE FIRST PRAYER BOOK OF KING EDWARD VI (1549)*[71] mainly because he believed it more closely reflected the faith and practice of the Church of Patristic times. The rubrics combined with the text of the service to make the Anglican office of matrimony the most comprehensive and practical instrument in all of Christendom. Wesley enjoyed conducting marriage ceremonies, and when he preached on the subject, he often cited passages from the marriage ritual.

The marriage ceremony of the Edwardian Prayer Book began with a rubric concerning the "banns." The term was originally Teutonic, meaning a public announcement that a particular couple intended to marry in the near future. The posting of the banns gave the public time to determine if the couple was "free to marry" or had some "impediment" that should prevent the union. At the marriage ceremony the priest would solemnly state, "Therefore if any man can shew any just cause why they may not lawfully be joined together, let him now speak, or else hereafter forever hold his peace." The rubric concerning the banns required that they be posted "three several Sundays or holydays, in the service time, the people being present" in the church.[72] If the couple came from two different parishes, the banns were required in both parishes. The marriage could be solemnized only in one of the two parishes, the officiating minister of that parish was required to possess a "certificate of the banns" issued by the minister of the other parish. This practice of the banns began in the 8th century and was still in force in the Church of England when John Wesley turned Sophy Hopkey Williamson away from the Savannah communion table for having violated the law by a sudden marriage without the banns.[73]

At the appointed time, in the church, the priest met the couple and their entourage and began the service. Edward's Prayer Book reads -

"Dearly beloved friends, we are gathered together here in the sight of God, and in the face of his congregation, to join together this man and this woman in holy matrimony; which is an honourable estate, instituted of God in Paradise, in the time of man's innocency, signifying unto us the mystical union that is betwixt Christ and his church; which holy estate Christ adorned and beautified with his presence, and first miracle that he wrought, in Cana of Galilee; and is commended of Saint Paul to be honourable among all men; and therefore is not to be enterprised, nor taken in hand unadvisedly, lightly, or wantonly, to satisfy men's carnal lusts and appetites, like brute beasts that have no understanding; but reverently, discreetly, advisedly, soberly, and in the fear of God; duly considering the causes for which matrimony was ordained. One cause was the procreation of children, to be brought up in the fear and nurture of the Lord, and praise of God. Secondly, it was ordained for a remedy against sin, and to avoid fornication; that such persons as be married might live chastely in matrimony, and keep themselves undefiled members of Christ's body. Thirdly, for the mutual society, help, and comfort, that the one ought to have of the other, both in prosperity and adversity. Into the which holy estate these two persons present come now to be joined. Therefore if any man can shew any just cause why they may not lawfully be joined so together, let him now speak, or else hereafter for ever hold his peace."[74]

The rubric then reads - "And also speaking to the persons that shall be married, he shall say," - and the priest was to admonish -

"I require and charge you, (as you will answer at the dreadful day of judgment, when the secrets of all hearts shall be disclosed,) that if either of you do know any impediment why ye may not be lawfully joined together in matrimony, that ye confess it. For be ye well assured, that so many as be coupled together otherwise than God's word doth allow, are not joined of God, neither is their matrimony lawful."[75]

The following rubric treated the occasion of someone in the congregation coming forward to state an objection to the marriage. Such a person was required to "be bound, and sureties with him, to the parties." If not bound, at least "put in a caution, to the full value of such charges as the persons to be married do sustain, to prove his allegation." Then the solemnization "must be deferred unto such time as the truth be tried." But, to the relief of all, when no allegation was made, the priest was allowed to continue his officiating, first asking the man -

"(Name) Wilt thou have this woman to thy wedded wife, to live together after God's ordinance in the holy estate of matrimony? Wilt thou love her, comfort her, honour and keep her in sickness and in health; and, forsaking all other, keep thee only to her so long as you both shall live?"[76]

The rubric reads - "The man shall answer, I will." Then it adds, "Then shall the Priest say to the woman,

(Name) Wilt thou have this man to thy wedded husband, to live together after God's ordinance in the holy estate of matrimony? Wilt thou obey him and serve him, love, honour, and keep him in sickness and in health; and forsaking all other, keep thee only to him, so long as you both shall live?"[77]

The rubric reads - "The woman shall answer, I will." The priest asks immediately thereafter, "Who giveth this woman to be married to his man?" The next rubric reads - "And the Ministers, receiving the woman at her father or friend's hands, shall cause the man to take the woman by the right hand, and so either to give their troth to other; the man first saying,

I (Name) take thee (Name) to my wedded wife, to have and to hold, from this day forward, for better for worse, for richer for poorer, in sickness and in health, to love and to cherish, till death us depart, according to God's holy ordinance; and thereto I plight thee my troth."[78]

The term "troth" is a very ancient term, meaning marital faith, fidelity and loyalty. The following rubric requires that "they shall loose their hands; and the woman, taking again the man by the right hand, shall say,

I (Name) take thee (Name) to my wedded husband, to have and to hold from this day forward, for better for worse, for richer for poorer, in sickness and in health, to love, cherish, and to obey, till death us depart, according to God's holy ordinance; and thereto I give thee my troth."[79]

Upon loosing hands, the rubric instructs the man to give the woman a ring and "other tokens of spousage, as gold or silver, laying the same upon the book." The priest delivers the ring to the man who places it upon the fourth finger of the bride's left hand. "And the man, taught by the Priest, shall say,

With this ring I thee wed, this gold and silver I thee give, with my body I thee worship, and with all my worldly goods I thee endow: in the Name of the Father, and of the Son, and of the Holy Ghost. Amen."[80]

The minister then prays,

"O ETERNAL GOD, Creator and Preserver of all mankind, Giver of all spiritual grace, the Author of everlasting life; Send thy blessing upon these they servants, this man and this woman, whom we bless in thy name; that as Isaac and Rebecca (after bracelets and jewels of gold given of the one to the other for tokens of their matrimony) lived faithfully together, so these persons may surely perform and keep the vow and covenant betwixt them made, whereof this ring given and received is a token and pledge, and may ever remain in perfect love and peace together, and live according to thy laws; through Jesus Christ our Lord. Amen."[81]

The minister, joining the couple's right hands, now declares, "Those whom God hath joined together let no man put asunder." The final pronouncement follows promptly -

"Forasmuch as (Name) and (Name) have consented together in holy wedlock, and have witnessed the same here before God and this company, and thereto have given and pledged their troth either to other, and have declared the same by giving and receiving gold and silver, and by joining of hands; I pronounce that they are man and wife together, in the name of the Father, of the Son, and of the Holy Ghost. Amen."[82]

A very ancient Christian blessing is added by the priest as a special benediction, featuring the Sign of the Cross upon the mention
of "God the Son." Wesley used the Sign of the Cross in the marriage ceremony since the Prayer Book called for it, and he also
used it in his baptismal services.[83] Following the blessing, the couple enter the quire and are seated while the minister reads or sings
one of the Psalms indicated in the Prayer Book - Psalm CXXVIII,
Beati omnes or Psalm LXVII, *Deus Misereatur nostri.* The *Gloria
Patri* then follows. The couple returns to the altar, and kneeling
they alternately pray the Lord's Prayer with the minister. He then
offers several prayers for them, after which a sermon on marriage
may follow with the Holy Communion being administered.[84]

Wesley's Methodist pilgrim must remember that the marriage
vow includes "till death us depart." No matter how unhappy or dis-
cordant the relationship may become, husband and wife have
vowed before God and the Christian community to live together in
the search for concord and mutuality until death comes to one or
the other. As for marriage after the resurrection, Wesley believed
that it does not survive death, that marriage is an institution of time
alone.[85] The only marriage experienced in the world to come is between Christ and his Church. His bride "is all holy men, the whole
invisible Church."[86]

ON THE SINGLE LIFE

Wesley, like St. Paul, recognized that not all pilgrims are to enter
the married state here in this life. Such persons are not forbidden
to marry, but they voluntarily choose the single life for the purpose
of greater devotion to Christ. Wesley wrote a tract upon this subject, entitled *THOUGHTS ON A SINGLE LIFE*, dated June 11,
1785. The arguments, however, are from Paul's discourse in I
Corinthians VII and Jesus's statement in Matthew XIX, 44ff, concerning "eunuchs for the kingdom of heaven." Wesley had long
held the single state as ideal, assuring his followers who married,
however, that they had not sinned by so doing.[87] From his Oxford

days he regarded the single state as representing greater devotion and opportunity for service to Christ.

At the outset of the *THOUGHTS ON A SINGLE LIFE*, Wesley rejected the Roman Catholic prohibition against clerical marriage as one of the "doctrines of devils." Likewise he attacked the affirmation of Medieval mystic authors who declared "marriage is only licensed fornication." While he saw many "troubles" attending the married state, Wesley believed it to be "honourable in all, and the bed undefiled." In fact, "persons may be as holy in a married as in a single state."[88] Nevertheless, as Wesley finally argued, such a thing is rare.

Quoting Paul, Wesley commented -

"We must not forget what the Apostle subjoins in the following verses: 'I say to the unmarried and widows, It is good for them, if they abide even as I. Art thou bound unto a wife? Seek not to be loosed. Art thou loosed for a wife? Seek not a wife. But if thou marry, thou hast not sinned. Nevertheless, such shall have trouble in the flesh. I would have you without carefulness. He that is unmarried careth for the things of the Lord, how he may please the Lord; but he that is married careth for the things of the world, how he may please his wife. The unmarried woman careth for the things of the Lord, that she may be holy both in body and spirit; but she that is married careth for the things of the world, how she may please her husband. And this I speak for your own profit, that you may attend upon the Lord without distraction.' (Verses 8, 27, 28, 32-35) 4. But though 'it is good for a man not to touch a woman,' (Verse 1,) yet this is not an universal rule. 'I would,' indeed, says the Apostle, 'that all men were as myself.' (Verse 7.) But that cannot be; for every man hath his proper gift of God, one after one manner, another after that.' 'If,' then, 'they cannot contain, let them marry; for it is better to marry than to burn.' (Verse 9.) 'To avoid fornication, let every man have his own wife, and let every woman have her own husband.' Exactly agreeable to this are the words of our Lord. When the Apostles said, 'If the case be so, it is good not to marry; he said unto them, All men cannot receive this saying, but they to whom it is given. For there are some eunuchs, who were so born from their mother's womb; there are some, who were made eunuchs by men; and there are eunuchs for the kingdom of heaven's sake. He that is able to receive it, let him receive it.' (Matthew XIX, 10-12.)"[89]

One may well ask Wesley, "Who is able to receive this saying and thus avoid marriage and burning?" The answer lies within an individual person, and no one should judge another in this matter. God gives to every person, at the moment when he or she is "first justified," the gift of grace for a single life - "but with most it does not continue long."[90] Its withdrawal is more probably the believer's fault than God's. New pilgrims should learn the advantages of maintaining the gift for the single life. The advantages, as Wesley enumerated them, are:

1. "You may be without carefulness" - free from caring for the things of the world. "You have only to care for the things of the Lord, how you may please the Lord. . .how you may be holy both in body and spirit."[91]

2. "You may attend upon the Lord without distraction" - some persons are like Martha, cumbered with much serving, "you may remain centered in God, sitting like Mary, at the Master's feet, and listening to every word of his mouth."[92]

3. "You may enjoy a blessed liberty from the 'trouble in the flesh'" - from "a thousand nameless domestic trials which are found, sooner or later, in every family." There are occasions of sorrow and anxiety, sickness, weakness, unhappy or disobedient children, wicked servants, and incorrigible sons and daughters.[93]

4. "You are at liberty from the greatest of all entanglements, the loving one creature above all others" - How difficult it is to give God our whole heart when one creature "has so large a share of it!" Wesley reasoned further, "How much more easily may we do this, when the heart is, tenderly indeed, but equally attached to more than one; or, at least, without any great inequality!"[94]

5. "You have leisure to improve yourself in every kind" - in relation to God and neighbour, while those who are married "are necessarily taken up with the things of the world." Wesley added,

"You may give all your time to God without interruption, and need ask leave of none but yourself so to do. You may employ every hour in what you judge to be the most excellent way. But if you was married, you may ask leave of your companion; otherwise what complaints or disgust would follow! And how hard it is even to know (how much more to act suitably to that knowledge) how far you ought to give way, for peace' sake, and where to stop! What wisdom is requisite, in order to know how far you can recede from what is most excellent, particularly with regard to conversation that is not 'to the use of edifying,' in order to please your good-natured or ill-natured partner, without displeasing God!"[95]

6. "You may give all your worldly substance to God" - If single, nothing can hinder you from doing this. "You have no increasing family. . .no wife. . .no children to provide for." Moreover, the single person will never be torn by doubt concerning whether he has done too much or too little in providing for a family.[96]

The advantages cited are the most obvious ones to the single pilgrim, according to Wesley. However, they may easily be lost. Consequently, Wesley exhorted his Methodist followers in the single state to: (1) "Prize the advantages you enjoy." (2) "Know the value of them." (3) "Pray constantly. . .let it be a matter of daily thanksgiving to God, that he has made you a partaker of these benefits." (4) "Be careful to keep them. . .so strong are the temptations which you will meet with to cast them away. Not only the children of the world, but the children of God, will undoubtedly tempt you thereto." (5) Converse "frequently and freely with those of your own sex who are likeminded." (6) "Avoid all needless conversation, much more all intimacy, with those of the opposite sex. . .unless you observe this, you will surely cast away the gift of God." In this matter, "Keep your heart with all diligence...Check the first risings of desire. Watch against every sally of imagination, particularly if it be pleasing." If the temptation persists, cry out, "My God and my all, I am thine, thine alone! I will be thine for ever! O Save me from setting up an idol in my heart! Save me from taking any step toward it. . .bring my every thought into captivity to the obedience of Christ." (7) "Avoid the sin of Onan" - that is, masturbation. (8) Avoid "all softness and effeminacy" for the Apostle declares that such persons "shall not inherit the kingdom of God." (9) "Avoid all delicacy, first in spirit, then in apparel, food, lodging, and a thousand nameless things." (10) "Avoid all needless self-indulgence, as well as delicacy and softness." (11) "Avoid all sloth, inactivity, indolence. Sleep no more than nature requires." 12) "Be never idle; and use as much bodily exercise as your strength will allow." (13) "Avoid all that pleasure which anyway hinders you enjoying him (God); yea, all such pleasure as does not prepare you for taking pleasure in God." (14) "Add to this constant and continued course of universal self-denial, the taking up of your cross daily, the enduring 'hardship as a good soldier of Jesus Christ.'" And (15) "Think not of a smoother path. Add to your other exercises constant and prudent fasting."[97]

The final exhortation represents Wesley's logical mind at work deductively, making explicit what resides implicitly in the previous exhortations - "I advise you. . .if you desire to keep them, use all the advantages you enjoy. Indeed, without this, it is utterly impossible to keep them; for the mouth of the Lord hath spoken the word which cannot be broken. . .with regard to all the good gifts of God." Those who live by God's gift for the single life are the spiritual eunuchs of the kingdom, "who abstain from things lawful. . .in order to be more devoted to God."[98] The words of Jesus form a special blessing for these persons - "There is no man that hath left . . .house, or brethren, or sisters, or father, or mother, or wife, or children, or lands, for my sake and the gospel's, but he shall receive an hundred fold now in this time; and in the world to come eternal life."[99]

The Christian pilgrim, then, may either marry or remain single. There is an appropriate gift from God for either. However, Wesley saw the single life as the more advantageous of the two. While his *THOUGHTS ON A SINGLE LIFE* was published in 1785, its position was not a newly formed one, embraced in doterage. An entry in the JOURNAL, dated November 5, 1764, traces his long indebtedness to the single life - "My scraps of time this week I employed in setting down my present thoughts upon a single life, which indeed, are just the same they have been these thirty years; and the same they must be, unless I give up my Bible."[100] "Thirty years" would push back the calendar to 1734, some time before he departed for Georgia. In fact, while at Oxford his Holy Club embraced voluntary celibacy, including in the Oxford Rules for

Methodists some rather familiar exhortations: "Avoid all manner of passion;" "Avoid. . .freedom with women;" and "Avoid the very beginning of lust."[101] At that same time, Wesley conducted self-examination sessions every Saturday night, always asking himself, "Have I loved women. . .more than God?"[102] Aboard ship for Georgia, he read A'Kempis with an enthusiasm unlike that when he first was introduced to these writings in 1725.[103] To a young priest seeking to save his soul by going to the New World, the words of Thomas A'Kempis must have burned deeply into Wesley's psyche -

"If thou wilt stand fast as thou ought, and grow in grace, esteem thyself as an exile and a stranger upon earth. Thou must be made a fool for Christ's sake, if thou desire to lead a religious life. The wearing of a religious habit, and the shaving of the crown, do little profit; but change of manners, and perfect mortification of passions, make a true religious man.

He that seeketh anything else but merely God, and the welfare of his own soul, shall find nothing but tribulation and sorrow. Neither can he stand long in peace, that laboureth not to be the least, and subject unto all.

Thou camest to serve, not to rule. Know that thou wast called to suffer and to labour, not to be idle, and spend thy time in talk. Here therefore men are proved as gold in the furnace. Here no man can stand, unless he be willing to humble himself with his whole heart for the love of God.

Gaze upon the lively examples of the holy Fathers, in whom true perfection and religion shined, and thou shalt see how little it is and almost nothing, which we do now in these days. Alas! what is our life, if it be compared with them?

The Saints and friends of Christ served the Lord in hunger and thirst, in cold and nakedness, in labour and weariness, in watchings and fastings, in prayer and holy meditations, in many persecutions and reproaches.

O how many and grievous tribulations did the Apostles, Martyrs, Confessors, Virgins, and all the rest suffer, that willed to follow the steps of Christ! For they hated their lives in this world, that they might keep them unto life eternal. O how strict and self-renouncing a life did those holy Fathers lead in the wilderness! How long and grievous temptations suffered they! How often were they assaulted by the enemy! What frequent and fervent prayers offered they to God! What rigorous abstinence did they fulfil! How great zeal and ardour had they for their spiritual progress! How fierce a war they waged for the taming of their faults! How pure and upright an intention kept they towards God! All riches, dignities, honours, friends, and kinsfolk they renounced, they desired to have nothing which appertaineth to the world; they scarce took things necessary for the sustenance of life; they grieved to serve their bodies even in necessity. Poor therefore were they in earthly things, but rich exceedingly in grace and virtues. Outwardly they were destitute, but inwardly they were refreshed with grace and divine consolation. To the world they were strangers, but near and familiar friends to God."[104]

Young Wesley felt closer to the kingdom of God as he read from A'Kempis, using select passages for daily devotions aboard ship,[105] and later at Savannah.[106] He struggled to take the advice of A'Kempis - *Omnes bonas mulieres devita, easque Deo commenda* (All good women avoid, and commend them to God).[107] He also knew the warning of A'Kempis - "Converse not much with young women!"[108] Unfortunately for Wesley, he failed to heed both bits of advice. Mrs. Hawkins nearly killed him and Miss Sophy Hopkey nearly landed him in marriage. Moreover, Wesley actively sought the company of women, and between 1725 and 1735 he conducted intense correspondence with several married women under the pen name of Cyrus. He wrote to Mrs. Mary Pendarves, whom he cryptically named "Aspasia." Mrs. Sally Kirkham became his "Varanese," and Mrs. Ann Granville became his "Selima."[109] As he grew older, the habit of writing to married women did not cease. One such letter had serious repercussions. On January 7, 1758, Wesley wrote to Mrs. Sarah Ryan. This was not the first time, but it was the wrong time, for Mrs. John Wesley intercepted the letter before it could be posted.[110]

WESLEY'S MARRIAGE

Wesley was constantly torn between the ideal of singleness and the need for a woman in his life. Most of his life was spent in the former state, during which time he spoke glowingly of the greater gift. When he decided to marry in February 1751, he said,

"Having received a full answer from Mr. Vincent Perronet, I was clearly convinced that I ought to marry. For many years I remained single, because I believed I could be more useful in a single than in a married state. And I praise God, who enabled me so to do. I now as fully believed that in my present circumstances I might be more useful in a married state; into which, upon this clear conviction, and by the advice of my friend, I entered a few days after."[111]

Wesley married on February 18, 1751, although he had made his decision earlier on the second of the month. On the sixth of February, he met with the single men of the London Society - "and (I) showed them on how many accounts it was good for those who had received that gift from God to remain 'single for the kingdom of heaven's sake;' unless where a particular case might be an exception to the general rule."[112]

In his excellent biography of John Wesley, Ingvar Haddal treats Wesley's love affairs with several women. Haddal refers to Miss Sophy Hopkey as "Wesley's Great Love," and Grace Murray as his "Last Love." Needless to say, Wesley married neither of these ladies. Haddal is correct in showing Wesley the loser in both instances - Miss Sophy and Grace suddenly married other suitors when Wesley dragged his feet.[113] Yet, Wesley told the story rather completely, if one is willing to dig it out of the *JOURNAL, DIARY,* and *LETTERS.*

The marriage of John Wesley - a fall from God's gift for singleness and from great and last love - brought many troubles to his flesh, mind, and soul. Despite this unfortunate marriage, Wesley's pilgrimage and ministry continued. None of his associates approved of his choice, although they had advised him to marry a woman of middle-age maturity and piety.[114] "The marriage irritated all concerned," Nehemiah Curnock observed, " - John, the bride, Ebenezer Blackwell, and the whole London Society."[115]

Charles Wesley, John's younger brother, put the developments of the marriage in chronological order. Writing in his *JOURNAL*, the younger Wesley recorded -

"Feb. 2, Sat. - My brother returned from Oxford, sent for and told me he was resolved to marry! I was thunderstruck, and could only answer he had given me the first blow, and his marriage would come like the *coup de grace*. Trusty Ned Perronet followed, and told me the person was Mrs. Vazeille - one of whom I never had the least suspicion. I refused his company (his brother's?) to the chapel (West Street) and retired to mourn with my faithful Sally. Groaned all the day, and several following ones, under my own and the people's burden. I could eat no pleasant food, nor preach, nor rest either by night or by day."[116]

John Wesley's intention to marry Mrs. Vazeille was attested to by Ned Perronet on February 2, leaving Charles Wesley in a dire state of depression. On February 10 following, John Wesley met with an accident on London Bridge. Slipping on the ice, he injured his ankle and was almost rendered immobile. Nevertheless, he went on to the Chapel at Snowfields and preached, seeking the aid of a surgeon to bind the ankle after the service. Later in the evening, finding it impossible to preach another time at the Foundery, Wesley requested that he be carried to Threadneedle Street.[117] There in the home of Mrs. Vazeille he found a week of tender and compassionate nursing.[118] Wesley did not tell the entire truth when he wrote in the JOURNAL about that week - "I spent the remainder of the week partly in prayer, reading, and conversation, partly in writing an *Hebrew Grammar* and *Lessons for Children*."[119] He also proposed marriage to his nurse. He had a romantic proclivity toward his nurses. His first nurse was Miss Sophy Hopkey in Georgia. His second nurse was Grace Murray who cared for him while he was ill in Newcastle.[120] Haddal observed, "On both occasions Wesley had felt the urge to marry his nurse, touched by her interest and care for his well-being."[121] However, it was his third nurse, Mrs. Vazeille, that he asked to become his wife, and she readily agreed to the proposal. He viewed his accident as an act of Providence, bringing him under the care of the woman he had earlier selected to become his wife. Edwards captures the urgency of Wesley's choice -

"He was forty eight years old and time would not wait indefinitely for him to make up his mind. As previously he had been too slow, so now he was too fast. Within the week he had not only proposed, but had married. At that time he could not easily set his foot to the ground and when he preached the following Sunday he did so in a kneeling position. The adage that those who marry in haste repent in leisure precisely applied to Wesley's action."[122]

Waxing philosophical, Bishop Francis J. McConnell observed, "A marriage contracted on the basis of the relation of nurse and sick man is always of doubtful worth."[123] McConnell argued further that it would have made little difference if Wesley had married his second nurse rather than his third - "He would have acted the same way with Mrs. Murray, as he did with Mrs. Vazeille. Probably Mrs. Murray might have been more submissive than Mrs. Vazeille, but that would likely have been the only difference."[124] Edwards was right when he declared that Wesley "was prepared to make no concessions because of his married state."[125] One month after the wedding, Wesley exclaimed to a friend, "I cannot understand how a Methodist preacher can answer it to God to preach one sermon or travel one day less in a married than in a single state."[126] The marriage was headed for serious difficulties.

The *GENTLEMAN'S MAGAZINE* announced the marriage to the world, citing the ceremonial day as February 18. The LONDON MAGAZINE also featured the occasion, but dated it on February 19, with the Reverend Mr. Charles Manning officiating. The bride was Mrs. Vazeille, the *LONDON MAGAZINE* reported, "the widow of a merchant, with a jointure of 10,000 pounds settled on herself and four children."[127] Her former husband was actually a banker.[128] In addition to owning the house on Threadneedle Street in London, the bride of Wesley owned a country house in Wandsworth.[129] When they married, Wesley "arranged for the whole of her fortune to be handed over to the children."[130] Edwards noted that Wesley invested the inheritance at 3% interest, and he never used a penny of it.[131]

When John and Molly Wesley appeared at the West Street Chapel on Wednesday, February 27, 1751 - better than a week after the wedding - John delivered an apology for his taking of Molly Vazeille as his wife. His younger brother, Charles, being in attendance, "ran away when he began his apology."[132] When he delivered the same apology at the Foundery, Charles reported, the members of the Society "hid their faces."[133] Why such consternation over the marriage? Nehemiah Curnock rightly observed that the *RULES OF THE HOLY CLUB* - drawn up by Wesley and his Oxford University peers years earlier - were still held as binding upon its members and the Methodist Societies. One of those rules amounted to the posting of banns before marriage, so each member of the Holy Club or Society could evaluate the forthcoming marriage. The consent of Club and Society was needed before a marriage could be solemnized.[134] Both George Whitefield and Charles Wesley complied with this rule, John Wesley consenting to their selections of partners. Frederick C. Gill correctly notes that John broke the agreement with his younger brother and the Societies. Charles Wesley lamented, "I was one of the last that heard of his unhappy marriage."[135] The ironic twist is that Wesley failed to post the banns in the Hays parish where he was married. Three Sundays of posted banns were required before solemnization could be performed. Wesley and the Reverend Mr. Manning disregarded this requirement, a requirement that Wesley once cited in reference to the marriage of Mrs. Sophy Hopkey Williamson. Family, friends, and Societies were greatly disturbed with the union. Wesley became defensive and gave his bold apology far and wide. The JOURNAL of Charles Wesley treats the nature of this apology -

"March 1, Fri., 1751 - Miss Hardy related my brother's apology: 'That in Oxford he had an independent fellowship, was universally honored, but left all for the people's sake (he is referring to this recent visit to the University); returned to town, took up his cross, and married; that at Oxford he had no more thought of a woman than for any other being; that he married to break down the prejudice about the world and him.' His easy-won lady sat openeyed. He said, 'I am not more sure that God sent His Son into the world than that it is His will I should marry.'"[136]

Family, friends, and Society members recognized that Wesley's temperament was not compatible with the married state, and that in spite of Mrs. Wesley's fine qualities, she was no congenial partner for a travelling evangelist.[137] Many studies of this subject have painted Molly Wesley as the source of marital failure - a most unfortunate interpretation, indeed! The state of holy matrimony requires a great degree of "cleaving" unto one's wife, but Wesley simply continued his extensive travels. At first he took her with him, but soon he went alone, as if he were single. In the Old Testament, the Torah commanded the husband to remain home with his wife for the first full year, exempting him from military duty and all other obligations. John Wesley felt himself under a higher law that subordinated his marriage to the itinerate ministry from the outset of pledging his troth to Molly. Ingvar Haddal puts this relationship into proper perspective -

"Her husband resisted the temptations of domestic ease so Mrs. Wesley fell in with his wishes and accompanied him on his travels round the country. Leaving her quiet and comfortable home, she lived through riots, street-fighting and stonings, she journeyed along muddy highways and snow-covered tracks, spent the night in filthy inns, served with food that was offensive to her housewifely instincts. She made an honest attempt to keep pace with this bundle of energy that she had taken as husband, but eventually it became too much for her."[138]

Writing to Ebenezer Blackwell, Molly's close friend, John related, in his letter of May 14, 1751, that his wife "has many trials ...If these encumbrances were removed, it might be a means of her spending more time with me."[139] Ned Perronet, Wesley's faithful confidant, wrote to Charles about Molly's and John's relationship - "I think the unhappy lady is most to be pitied, though the gentleman's case is mournful enough."[140]

The mutual estate of misery increased in volume, but, to his credit, Wesley never struck his wife. After all, he had preached and taught, there is no biblical justification for a husband striking his wife. Nor is there any scriptural authority for a wife to beat her husband, although on one occasion Molly was accused of doing so.

The episode was told by John Hampson, one of Wesley's preachers who later left the Methodists -

"I was once on the point of committing murder. Once, when I was in the north of Ireland, I went into a room, and found Mrs. Wesley foaming with fury. Her husband was on the floor, where she had been trailing him by the hair of his head; and she herself was still holding in her hand venerable locks which she had plucked up by the roots. I felt as though I could have knocked the soul out of her."[141]

His waking and sleeping hours were haunted by the dynamic tensions of marriage. In a letter to his wife, dated May 7, 1756, Wesley implored - "I dreamed last night that I was carried to execution and had but a few minutes to live ...While we live, let us love; that if we do not meet again here, we may in a better place. My dear Molly, Adieu."[142] The tensions increased. On January 27, 1758, Wesley confided by letter to Mrs. Sarah Ryan, "My Dear Sister. Last Friday, after many severe words, my wife left me, vowing she would see me no more." In the remainder of the letter, Wesley recounted how Molly discovered and read his unsealed letter intended for Mrs. Ryan.[143] A reconciliation was effected, but the tensions continued unabated. Two days before Christmas in 1758, being away on his rounds of Methodist Societies, he wrote to Molly - "I was much concerned, the night before I left London at your unkind and unjust accusation." Molly accused him of "unkindness, cruelty, and what not!" Why? Because he insisted on choosing his own company, and for speaking or writing to anyone he thought proper. He responded, "For more than seven years this has been a bone of contention between you and me. And it is so still." With determination he asserted, "For I will not, I cannot, I dare not give it up. . .but 'you will say all manner of evil of me.' Be it so; but still I stand just where I was. Then 'you will show my private letters to all the world.' If you do, I must assert my right still." His "right" meant his selection of companions and his conversing by speaking or writing to whomever he wished. His closing argument, a warning, places his ministry above his marriage and his wife's claim to his body, time, and affections - "O do not continue to trouble yourself and me and to disturb the children of God (Societies) by still grasping at a power which must be denied you."[144]

The warning settled nothing. Molly Wesley stole his letters as she had threatened. Ebenezer Blackwell was confident that he could persuade her to return them. Wesley wrote to Blackwell about this matter -

"My wife picks my lock and steals my papers. Afterwards she says, 'You cannot trust me.' I answer, 'I cannot till you restore what you stole and promise to steal no more.' She replies, 'I will burn them, or lodge them with another, on such terms.' I answer nothing...Permit me to add one word to you. You think yourself a match for her; but you are not."[145]

Blackwell then advised Wesley to return home and secure a locksmith to open Molly's bureau and take back his papers. Molly hastily surrendered his papers before he could carry out this plan. But, shortly thereafter, Molly "robbed" him again of the papers, and she made them available to about twenty different persons for the purpose of making them think ill of her husband. He was several hundred miles away at the time.[146]

No Christian Pilgrim could have had as much trouble in the flesh as Wesley had by October 1759. Writing to Molly, he listed ten "dislikes" in their relationship -

"I will tell you simply and plainly the things which I dislike - 1. showing anyone my letters and private papers. 2. not having the command of my own house. 3. being a prisoner in my own house. 4. being a prisoner at large, even when I go abroad. 5. not being safe in my own house - I cannot call even my study, even my bureau, my own. They are liable to be plundered every day. . . I miss money too, and he that will steal a pin will steal a pound. 6. your treatment of my servants. . . you browbeat, harass, rate them like dogs, make them afraid to speak to me. . . you forget even good breeding, and use such coarse language as befits none but a fishwife. 7. your talking against me behind my back, and that every day and almost every hour of the day. 8. your slandering of me, laying to my charge things which you know are false - that I beat you, etc. 9. your common custom of saying things not true. 10. your extreme, immeasurable bitterness to all who endeavour to defend my character (as my brother, etc.), breaking out into foul, unmannerly language, such as ought not to defile a gentlewoman's lips if she did not believe one word of the Bible."[147]

By the following July (1760), Wesley was in the mood for a meaningful reconciliation. He wrote to Molly, appealing to her to "walk hand in hand, to Immanuel's land." "The evening of life" was at hand. "Yesterday is past, and not to be recalled; tomorrow is not ours. Now, Molly let us set out; 'Let us walk hand in hand, to Immanuel's land!' - Your affectionate husband." The next eleven years were tension filled, with few marital joys. On January 23, 1771, Molly left her husband for good.

Three years later, as Mr. Walsh laid dying, he asked Wesley, "How did you part with Mrs. Wesley the last time?" Wesley responded, "Very affectionately." Walsh exclaimed, "Why, what a woman is this! She told me your parting words were, 'I hope to see your wicked face no more?'"[148]

On October 8, 1781, while Wesley was on a mission, Molly Wesley died in London and was buried in the churchyard of Camberwell. When he returned to London the following Friday, he was told of her death and burial. The *GENTLEMAN'S MAGAZINE* gave a brief obituary notice - "Died Mrs. M. Wesley, aged 71, wife of Mr. John Wesley, the celebrated Methodist."[149]

Frederick C. Gill has judged Wesley's marriage rightly - "John paid a heavy price for his rashness in what proved to be the biggest blunder of his life."[150] The union of two strong-willed persons brought grief and despair rather than the biblical ideal of one flesh. Not only did these two persons suffer but so did their friends. As it finally turned out, Haddal observes, "His real marriage was to his work and in this he never faltered."[151] Edwards adds an interesting postscript to the death of Molly Wesley - "Her money went to a son and one ring went to her husband. She had the last word after all!"[152]

Wesley's venture from singleness into marriage, lasting for thirty years, never realized the ideal of biblical oneness, as he understood and taught it. Nevertheless, he continued to speak of the joys of Christian marriage as well as the challenges, of its rewards as well as the duties, and of the gift of grace needed by husbands and wives to bless their efforts at attaining the ideal.

CHILDREN OF THE MARRIAGE

It will be remembered that John Wesley viewed procreation as one of the basic functions of marriage. Children are an heritage of the Lord. Being the fruit of the mother's womb, they are God's reward, not the result "of man's own labour."[153] It is, moreover, the parents' duty to raise them up as godly members of God's kingdom. In his sermon ON FAMILY RELIGION, Wesley stated the case accordingly -

"Your children (are) immortal spirits whom God hath, for a time, entrusted to your care, that you may train them up in all holiness, and fit them for the enjoyment of God in eternity. This is a glorious and important trust; seeing one soul is of more value than all the world beside. Every child, therefore, you are to watch over with the utmost care, that, when you are called to give an account of each to the Father of Spirits, you may give your accounts with joy and not with grief."[154]

The children of Christian pilgrims are to be trained from birth for the "enjoyment of God" in eternity. Wesley's view of parental stewardship of children followed the covenantal theology of Richard Baxter, the famous Seventeenth Century Puritan divine. Baxter, arguing that children born of Christian believers are members of God's covenant through Christian circumcision (infant baptism), insisted that parents engage in the most strenuous education of their children to bring them at last to heaven.[155] But Wesley was indebted to other forms of covenant theology as well as Baxter's. The Bucerian phase of the reform of the Church of England, the Genevan phase wrought by the return of the Marian exiles upon the coronation of Queen Elizabeth I, the rise of the Arminian movement within the late Seventeenth Century Anglican Church, these all shared a common theology of the children of believers being in covenant with God by virtue of birth and the new birth of infant baptism. In all contributing traditions, such children are to be "nurtured" to a fiducial commitment at the age of discretion.

Christian nurture is the joint operation of the Holy Spirit and faith community, attempting to bring the baptized child to faith in Christ. It has two main objectives: (1) To preserve the child in baptismal grace; and (2) to prepare him through religious education for future faith. As such, nurture is the divine means of giving fulfillment to Christian baptism. It is a developmental process, making baptized children genuinely pious.[156] The personal religion produced by nurture brings the child an assurance of saving faith, a legitimate goal for them to pursue.[157] Nurture, therefore, is more than merely raising children so that they do no harm or abstain from outward sin. Moreover, it is more than intellectual training in theological doctrines. Its overall *telos* is to install in baptized children true religion, holiness, and the love of God and mankind - in short, the restoration of the moral image of God.[158] Wesley was convinced that nurture is the "ordinary means" by which Christians are made. In his letter to Miss Bishop, dated May 17, 1781, reminding her of the great task of the Christian educator, Wesley commanded her to work harder among her children and to "make Christians, my dear sister, make Christians!"[159]

His principle of nurture was based on four presuppositions. By themselves they seem eccentric. However, as John W. Prince observed, if Wesley's general theological premises are granted, they are rational and logical.[160]

A. At birth the mind is a *tabula rasa*, having no innate ideas.[161] Echoing John Locke, Wesley held that all knowledge is acquired gradually, by sensation and reflection.[162] Contrary to John Calvin's famous argument in his *INSTITUTES OF THE CHRISTIAN RELIGION*, there is no idea of God stamped upon the infant's mind at birth.[163] God can only be apprehended progressively through revelation and the empirical observation of his works.[164] Even then, however, godly teachers are essential for epistemological acquisition.[165]

B. Sin, as an inherent force, still resides in the soul of the infant following baptism. Wesley's understanding of sin was Augustinian, having two dynamic dimensions - guilt and a propensity towards evil. Wesley's view of infant baptism was also Augustinian in relation to the problem of sin. In baptism the infant of believing parents is justified, being washed of the guilt of original sin and given a new relationship with God. But the washing of baptism effects an element of regeneration - the new birth - which not only produces the divine life within the soul, but also produces the process of mortification by which the "propensity towards evil" begins gradually to die. Regeneration, therefore, brings an actual change to the nature of the infant. The "propensity" or "power" of sin does not have dominion over the baptized, and yet, it is not entirely destroyed. Its last vestiges are to be destroyed by the process and experience of sanctification.[166] Wesley called this remaining power of sin "the bias of nature." In baptized children it shows itself as self-will. Nurture, as progressive sanctification, effects the curing of this bias.[167]

C. Religious education is God's instrument in nurture for the rectifying of the "bias of nature." In his tract, *A THOUGHT ON THE MANNER OF EDUCATION CHILDREN* (1783), the case is stated quite strongly -

"Scripture, reason, and experience jointly testify that, inasmuch as the corruption of nature is earlier than our instruction can be, we should take all pains and care to counteract this corruption as early as possible. The bias of nature is set the wrong way: education is designed to set it right. This, by the grace of God, is to turn the bias from self-will, pride, anger, revenge, and the love of the world, to resignation, lowliness, meekness, and the love of God."[168]

Behind this presupposition is clearly the educational theory of John Locke (1632-1704). In his treatise, SOME THOUGHTS CONCERNING EDUCATION, Locke spoke of children born with "some byass in their natural temper, which it is the business of education to take off."[169] Wesley, like his mother, knew the works of Locke with an impeccable precision. What Lockean doctrines his mother did not teach him, Wesley learned at Christ Church College (Oxford University) where earlier Locke took his B.A. degree and gained scholarly reputation.

For Locke and Wesley, education must create two positive habits in baptized children - (1) that of constant submission of the will to the greater will of the parent; and (2) that of constant self-denial. The successful creation of these habits insures the curing of the "bias of nature."[170] For a more complete treatment of these habits, please consult my book, *FROM FONT TO FAITH: JOHN WESLEY ON INFANT BAPTISM AND THE NURTURE OF CHILDREN.*

D. The age of discretion is a stage of childhood when individual reason is sufficient enough to make choices and to be held responsible for the consequences. Hence, it is the time when an intelligent faith consent to the grace of God is possible. The High Church Tradition fixed the age at six or seven years.

When Wesley was eight, his father admitted him to the Lord's Supper for the first time, and Wesley never hesitated to admit children of six years old to the sacred table.[171] If nurture is effective, when discretion arrives, a faith consent will confirm the grace received in baptism, completing the baptismal process form font to personal faith.

The *FIRST PRAYER BOOK OF KING EDWARD VI*, which Wesley preferred because it more closely reflected the practices of the Early Church, required that young children, following their baptism, be taught "what their Godfathers and Godmothers promised for them in baptism."[172] In addition to mastering the meaning of their baptism, children were to master the Catechism and its treatment of the Ten Commandments, the Apostles' Creed, and the Lord's Prayer. The Catechism of the Church of England was not too different from other Protestant catechisms - Luther's *Kleiner Katechismus*, Calvin's *Genevan Catechism*, the *Heidelberg Catechism* and the *Westminster Catechism*. Wesley provided a catechism for the children of Methodist believers. He called it *INSTRUCTIONS FOR CHILDREN*. The work was a translation of a French book by Abbe Fleury and M. Pierre Poiret, entitled *Les Principes Solides De La Religion Et De La Vie Chretienne Appliques A L'Education Des-Enfants.*[173] Nevertheless, Wesley printed the translation as if it were his own composition, circulating it far and wide among his Methodists.

Wesley's plan for the complete nurturing of baptized children included intense religious education at home, in the Methodist societies, and in special Methodist schools begun for that purpose. He provided other printed materials to be used in this grand design - *TOKENS FOR CHILDREN*, an anthology containing the religious experience of young children; *LESSONS FOR CHILDREN*, a book of Bible studies graded for childhood stages; *EXPLANATORY NOTES UPON THE NEW TESTAMENT*, a commentary for older children and youth, as well as adults; *HYMNS FOR CHILDREN*; and *PRAYERS FOR CHILDREN*.

The "holy child" of the covenant, then, is also a pilgrim, on the way of grace, from birth through baptism to discretion and faith. As a child, he must be taught holy concepts, and he must be established in holy habits. Above all, he must pray for success in his quest for fiducial commitment -

"Bless to me thy word, o my Heavenly Father, and all the means of grace, that I may not use them in vain or to my hurt, but for the instruction of my mind, reforming my life, and the saving (of) my soul."[174]

"Be thou pleased to second every word of instruction that I have received with the power of thy grace and Holy Spirit. . .do thou give me a heart filled with thy love, and lifted up in thy praise, and devoted to thy honour and glory all the days of my life."[175]

"Grant me such grace, that I may be able to withstand the temptations of the world, the flesh, and the devil, and with a pure heart and mind to follow the steps of my gracious redeemer."[176]

"O let me be devoted to thee from my childhood. Keep out of my heart all love or the world, of riches, or any other created thing, and fill it with the love of God."[177]

Notes

1   WORKS, *op.cit.*, XIII, 43; VI, 452.
2   METHODIST HYMNAL, *op.cit.*, 801.
3   *Ibid*, 790.
4   EXPLANATORY NOTES OLD TESTAMENT, *op.cit.*, I, 13.
5   *Loc.cit.*
6   *Ibid*, I, 18.
7   *Loc.cit.*
8   *Ibid*, I, 20.
9   *Ibid*, I, 23. Cf. WORKS, *op.cit.*, V, 280.
10   *Ibid*, I, 13.
11   *Ibid*, I, 80.
12   *Ibid*, I, 887.   Cf. Ruth IV, 7ff.
13   *Ibid*, III, 2472.
14   EXPLANATORY NOTES NEW TESTAMENT, *op.cit.*, 202.
15   *Loc.cit.*
16   EXPLANATORY NOTES OLD TESTAMENT, *op.cit.*, I, 20, 843; II, 877.
17   *Ibid*, I, 25-26.
18   *Ibid*, I, 96.
19   EXPLANATORY NOTES NEW TESTAMENT, *op.cit.*, 280.
20   EXPLANATORY NOTES OLD TESTAMENT, *op.cit.*, I, 655f.
21   *Ibid*, I, 268.
22   *Ibid*, I, 601.
23   EXPLANATORY NOTES NEW TESTAMENT, *op.cit.*, 336.
24   EXPLANATORY NOTES OLD TESTAMENT, *op.cit.*, I, 656.
25   EXPLANATORY NOTES NEW TESTAMENT, *op.cit.*, 33.
26   *Ibid*, 92-93.
27   *Loc.cit.* Cf. *Ibid*, 173.
28   WORKS, *op.cit.*, V, 280.
29   *Loc.cit.*
30   EXPLANATORY NOTES OLD TESTAMENT, *op.cit.*, I, 656.
31   EXPLANATORY NOTES NEW TESTAMENT, *op.cit.*, 718.
32   EXPLANATORY NOTES OLD TESTAMENT, *op.cit.*, I, 396.
33   EXPLANATORY NOTES NEW TESTAMENT, *op.cit.*, 208.
34   *Ibid*, 718-719.
35   WORKS, *op.cit.*, VII, 80.
36   EXPLANATORY NOTES NEW TESTAMENT, *op.cit.*, 719.
37   *Loc.cit.*
38   *Loc.cit.*
39   *Ibid,*, 603.

40 *Loc.cit.*
41 *Loc.cit.*
42 *Loc.cit.*
43 *Ibid*, 604.
44 *Loc.cit.* Cf. WORKS, *op.cit.*, VI, 456-457. Wesley, following St. Paul, argued that no believer should enter into marriage with an unbeliever.
45 *Loc.cit.* Cf. Beveridge, *op.cit.*, 463. Wesley freely quoted from Beveridge here - "If their parents be heathens, the children are heathens; if their parents be Christian, the children are Christian too."
46 EXPLANATORY NOTES NEW TESTAMENT, *op.cit.*, 781-782.
47 WORKS, *op.cit.*, XIV, 231. Wesley added this work to his CHRISTIAN LIBRARY.
48 William Whateley, *A BRIDEBUSH OR A WEDDING SERMON* (London: William Iaggard, 1617).
49 *Ibid*, 1.
50 *Ibid*, 2.
51 Wiliam Whateley, *A CARE CLOTH OR A TREATISE OF THE CUMBERS AND TROUBLES OF MARRIAGE* (London: Felix Kyngston, 1624), 2-3.
52 *HOMILIES OF THE CHURCH OF ENGLAND* (London: SPCK, 1843). Cf. WORKS, *op.cit.*, I, 164, 224, 254; III, 30; V, 661, 239; VII, 204; VIII, 23, 31, 54, 55, 74, 103-105, 129, 130, 473, 474.
53 WORKS, *op.cit.*, V, 239.
54 HOMILIES (1843), *op.cit.*, 534-535.
55 *Ibid*, 535-536.
56 *Ibid*, 536.
57 WORKS, *op.cit.*, VII, 80.
58 HOMILIES (1843), *op.cit.*, 537.
59 *Loc.cit.*
60 *Ibid*, 540 Cf. EXPLANATORY NOTES NEW TESTAMENT, *op.cit.*, 880.
61 *Ibid*, 542.
62 *Ibid*, 544-545.
63 *Ibid*, 545.
64 *Ibid*, 547.
65 *Ibid*, 548.
66 WORKS, *op.cit.*, VI, 456.
67 *Ibid*, VI, 457. Cf. EXPLANATORY NOTES NEW TESTAMENT, *op.cit.*, 660.
68 *Ibid*, VI, 459-460.
69 *Ibid*, VII, 302.
70 *Ibid*, VI, 85.
71 JOURNAL, *op.cit.*, I, 210 note.
72 Edward Cardwell (Ed), *THE TWO BOOKS OF COMMON PRAYER* (Oxford: The University Press, 1841), 351.
73 JOURNAL, *op.cit.*, I, 337. Cf. Maldwyn Edwards, *MY DEAR SISTER* (Manchester: Penworth Ltd.), 31. Edwards commented - "So careless and

informal a wedding would hardly seem legal to Wesley and his appeal to the Bishop of London's Commissionary bore fruit because future marriages without banns were forbidden."

74 Cardwell, *op.cit.*, 351-352.

75 *Ibid*, 352. Cf. JOURNAL, *op.cit.*, I, 337. Wesley regarded Sophy Williamson's marriage invalid because of the impediment of not posting the required banns.

76 *Loc.cit.*

77 *Ibid*, 352-353.

78 *Ibid*, 353.

79 *Loc.cit.*

80 *Loc.cit.*

81 *Ibid*, 354.

82 *Loc.cit.*

83 John Wesley, *THE SUNDAY SERVICE OF THE METHODISTS* (London: 1788), 145.

84 Cardwell, *op.cit.*, 354-359.

85 EXPLANATORY NOTES NEW TESTAMENT, *op.cit.*, 106.

86 *Ibid*, 1033-1034.

87 *Ibid*, 606.

88 WORKS, *op.cit.*, XI, 457.

89 EXPLANTORY NOTES NEW TESTAMENT, *op.cit.*, 93, 603, 604, 606.

90 *Loc.cit.*

91 *Loc.cit.*

92 *Loc.cit.*

93 *Loc.cit.*

94 *Ibid*, 459.

95 *Loc.cit.*

96 *Loc.cit.*

97 *Ibid*, 459-462.

98 *Ibid*, 462.

99 *Ibid*, 463. Cf. JOURNAL, *op.cit.*, I, 373 - Wesley asked Spangenberg, "Is celibacy a state more advantageous for holiness than marriage?" Answer - "Yes, to them who are able to receive it."

100 JOURNAL, *op.cit.*, V, 101.

101 *Ibid*, I, 48.

102 *Ibid*, I, 52.

103 *Ibid*, I, 122. Cf. Baker, *op.cit.*, XXV, 162.

104 Thomas A'Kempis, *THE IMITATION OF CHRIST* (New York: Books, Inc.), 22-24. Cf. Lee, *op.cit.*, 26, 59, 79, 100, 145, 176-178, 198, 213-214, 225-226. Also, WORKS, *op.cit.*, VII, 422-424.

105 JOURNAL, *op.cit.*, I, 122, 124; III, 159.

106 *Ibid*, I, 182, 192, 198, 281, 318.

107 *Ibid*, I, 152.

108 *Ibid*, I, 153.

109   Baker, *op.cit.*, XXV, 246. Cf. JOURNAL, *op.cit.*, I, 67-69. Also, Ingvar Haddal, *JOHN WESLEY* (New York: Abingdon Press, 1960), 59 - ". . . women. . .listened to him."

110   Telford, *op.cit.*, IV, 4-5. Cf. WORKS, *op.cit.*, XII, 216-222.

111   JOURNAL, *op.cit.*, III, 512. On February 3, 1751, Wesley preached at Hayes upon Genesis I, 27 - "So God created man in his own image...male and female created him them." Cf. Frederick C. Gill, *CHARLES WESLEY THE FIRST METHODIST* (New York: Abingdon Press, 1969), 149. Gill says John Wesley married because of "his grief over Grace Murray."

112   *Loc.cit.*

113   Haddal, *op.cit.*, chapters IX and X. Cf. Edwards, DEAR SISTER, *op.cit.*, 38. Haddal borrowed from Legev's *WESLEY'S LAST LOVE* (1910), referring to Grace Murray.

114   JOURNAL, *op.cit.*, III, 514. Cf. Haddal, *op.cit.*, 142.

115   *Loc.cit.*

116   *Loc.cit.*

117   *Ibid*, III, 512-513.

118   Haddal, *op.cit.*, 143. Cf. DEAR SISTER, *op.cit.*, 42.

119   JOURNAL, *op.cit.*, III, 513.

120   Haddal, *op.cit.*, 143. Cf. DEAR SISTER, *op.cit.*, 33.

121   *Loc.cit.*

122   DEAR SISTER, *op.cit.*, 42.

123   McConnell, *op.cit.*, 225.

124   *Ibid*, 224.

125   DEAR SISTER, *op.cit.*, 42.

126   JOURNAL, *op.cit.*, III, 517.

127   *Ibid*, III, 513.

128   Gill, *op.cit.*, 150.

129   *Loc.cit.*

130   Haddal, *op.cit.*, 143.

131   DEAR SISTER, *op.cit.*, 39.

132   JOURNAL, *op.cit.*, III, 515.

133   *Loc.cit.*

134   *Loc.cit.*

135   Gill, *op.cit.*, 150.

136   JOURNAL, *op.cit.*, III, 515.

137   Gill, *op.cit.*, 150.

138   Haddal, *op.cit.*, 145.

139   WORKS, *op.cit.*, XII, 175.

140   Gill, *op.cit.*, 151.

141   Haddal, *op.cit.*, 146-147. Cf. Gill, *op.cit.*, 151.

142   Telford, *op.cit.*, III, 177.

143   *Ibid*, IV, 4-5.

144   *Ibid*, IV, 49-50.

145   *Ibid*, IV, 52-53. This letter is dated March 2, 1759.

146   *Ibid*, IV, 61-62. This material is from Wesley's letter to Molly, April 6, 1759.

147   *Ibid*, IV, 75-78.
148   *Ibid*, IV, 98-101.
149   JOURNAL, *op.cit.*, VI, 337. Cf. Telford, *op.cit.*, VII, 323 - Wesley commented on Samuel Bradburn, who became a widower in 1786, "I believe the loss of his wife will be one of the greatest blessings which he has ever met with in his life."
150   Gill, *op.cit.*, 151.
151   Haddal, *op.cit.*, 147.
152   DEAR SISTER, *op.cit.*, 44. Cf. Telford, *op.cit.*, VIII, 152. In 1790, Wesley wrote - "I married because I needed a home, in order to recover my health; and I did recover it. But I did not seek happiness thereby, and I did not find it."
153   EXPLANATORY NOTES OLD TESTAMENT, *op.cit.*, III, 1810.
154   WORKS, *op.cit.*, VII, 79.
155   Richard Baxter, *THE PRACTICAL WORKS OF RICHARD BAXTER* (London: Arthur Hall and Company, 1847), IV, 108. Cf. JOURNAL, *op.cit.*, VI, 218. Wesley cited the case of an adult who began sainthood as a very young child.
156   WORKS, *op.cit.*, XIII, 475ff.
157   John W. Prince, *JOHN WESLEY ON RELIGIOUS EDUCATION* (New York: Methodist Book Concern, 1926), 87.
158   WORKS, *op.cit.*, XIII, 475-476. Cf. Prince, *op.cit.*, 87-88.
159   *Ibid*, XIII, 37.
160   Prince, *op.cit.*, 148.
161   WORKS, *op.cit.*, XIII, 455.
162   *Ibid*, XIII, 456.
163   *Ibid*, VII, 91.
164   Cannon, *op.cit.*, 158. Cf. WORKS, *op.cit.*, VII, 82.
165   Alfred H. Body, *JOHN WESLEY AND EDUCATION* (London: The Epworth Press, 1936), 49.
166   Prince, *op.cit.*, 95.
167   WORKS, *op.cit.*, XIII, 476. Cf. Cannon, *op.cit.*, 93.
168   *Loc.cit.*
169   John Locke, *SOME THOUGHTS CONCERNING EDUCATION* (Cambridge: The University Press, 1913), 119.
170   Body, *op.cit.*, 56.
171   H. W. Holden, *JOHN WESLEY IN COMPANY WITH HIGH CHURCH-MEN* (London: The Church Press Company, 1870), 27-28.
172   *THE FIRST PRAYER BOOK OF KING EDWARD VI* (London: Griffith, Farran, Brown and Company, Ltd., 1911), 228. Cf. WORKS, *op.cit.*, X, 506.
173   Prince, *op.cit.*, 127.
174   WORKS, *op.cit.*, XI, 260.
175   *Ibid*, XI, 261.
176   *Ibid*, XI, 262.
177   *Ibid*, XI, 270.

A general footnote: In 1779, Wesley published his poem, *THE BACHELOR'S WISH* - containing many stanzas - in the *ARMINIAN MAGAZINE.* A few select stanzas follow -

A beautiful face let others prize,
The features of the fair,
I look for spirit in her eyes,
And meaning in her air.

O could I such a female find,
Such treasure in a wife,
I'd pass my days to peace resign'd
Nor fear the ills of life.

Chapter Eight

The Pilgrimage from Paradise Lost to Paradise Improved: The Pilgrim and Death

"In the day Adam ate forbidden fruit, he became mortal, he began to die; his whole life after was but a forfeited condemned life, nay it was a wasting dying life; he was not only like a criminal sentenced, but as one crucified, that dies slowly and by degrees."[1]

Since the Fall of our first parents, Wesley believed, death comes for us all. It comes for kings and peasants without any regard for status. It comes for believers and unbelievers. All humans are born mortal, entering a flawed life that progressively moves each person toward the grave, slowly and by degrees. While the book of Genesis relates the imposition of mortality upon the first parents, the New Testament Epistle to the Romans reveals the extent of mortality -

"Therefore as by one man sin entered into the world, and death by sin; even so death passed upon all men, in that all sinned. . . Death reigned from Adam to Moses, even over them that had not sinned after the likeness of Adam's transgression, who is the figure of him that was to come."[2]

Wesley's exegesis of this passage from Romans V is crucial to a proper appreciation of death in the Christian pilgrimage. It is not an enemy as many have contended, but a consequence of sin and subject to the transforming grace of the Second Adam, Jesus Christ, into eternal life. But, let us permit Wesley to relate this soteriological perspective.

In his *EXPLANATORY NOTES UPON THE NEW TESTA-MENT*, Wesley explained how Adam's first sin resulted in a universal introduction of actual sin, "and its consequence, a sinful nature. And death - with all its attendants."[3] Of such a tragic sin Wesley exclaimed -

> "It entered into the world when it entered into being; for till then it did not exist. . . All sinned in Adam. These words assign the reason why death came upon all men; infants themselves not excepted, in that all sinned. . . Death reigned - and how vast is his kingdom! Scarce can we find any king who has as many subjects, as are the kings whom he hath conquered. Even over them that had not sinned after the likeness of Adam's transgression - Even over infants who had never sinned, as Adam did, in their own persons; and over others who had not, like him, sinned against an express law."[4]

Following the argument of St. Paul, Wesley made a contrast between Adam and Christ -

> "Who is the figure of him that was to come - Each of them being a public person, and a federal head of mankind. The one, the fountain of sin and death to mankind by his offence; the other, of righteousness and life by his free gift. Thus far the apostle shows the agreement between the first and second Adam; afterwards he shows the differences between them. The agreement may be summed up thus: As by one man sin entered into the world, and death by sin; so by one man righteousness entered into the world, and life by righteousness. As death passed upon all men, in that all had sinned; so life passed upon all men (who are in the second Adam by faith), in that all are justified. And as death through the sin of the first Adam reigned even over them who had not sinned after the likeness of Adam's transgression; so through the righteousness of Christ, even those who have not obeyed, after the likeness of His obedience, shall reign in life."[5]

There is a clear distinction to be made, according to Wesley, between "grace" and the "gift" of Jesus Christ, the second Adam. Adam's offence is opposed by grace. Death, the consequence of the offence, is opposed by the gift of Christ, which is the "gift of life."[6] St. Paul called the imparting of this gift "Justification to life."

Wesley added, ". . .that sentence of God, by which a sinner under sentence of death is adjudged to life."[7]

Christ's gift to life was prophesied in the Old Testament by the 8th century prophet Hosea, Wesley maintained. The text reads, "I will ransom them from the power of the grave; I will redeem them from death: O death, I will be thy plagues, O grave, I will be thy destruction."[8] Wesley's commentary makes Christ the fulfillment of this prophecy -

> *Ransom* - By power and purchase by the blood of the lamb of God, and by the power of his Godhead. *Them* - That repent and believe. *From the grave* - He conquered the grave, and will at the great day of resurrection open those prison doors, and bring us out in glory. *From death* - and from the second death, which shall have no power over us. *Thy plagues* - Thus I will destroy death. I will pull down those prison walls, and bring out all that are confined therein, the bad of whom I will remove into other prisons, the good I will restore to glorious liberty."[9]

Jesus Christ, in his death, burial, and resurrection, has ratified this prophetic promise. Wesley argued that Christian pilgrims move through these earthly days waiting for God's summons for their souls to leave their bodies - "Leaving the old, both worlds at once they view, Who stand upon the threshold of the new!"[10] In this transport, the pilgrims can look both backwards and forwards - if back to the earthly life, there is the "calm remembrance of a life well spent" - if forwards into eternity, "there is an inheritance incorruptible, undefiled, and that fadeth not away." Moreover, in the moment of death the pilgrim sees "a convoy of angels ready to carry him into Abraham's bosom." Wesley referred to these angels as "the ministering host of invisible friends."[11] If the pilgrim can expect such a convoy of friends at death, how can death be feared?

As stated earlier in this work, Wesley relied heavily upon the HOMILIES of the Church of England for theological clarity, and he accepted these standard sermons as normative for Methodist theology. The *HOMILY ON THE FEAR OF DEATH*, well known by Wesley from his early childhood, states most masterfully -

"Thus is this bodily death a door or entering unto life, and therefore not so much dreadful. . .as it is comfortable; not a mischief, but a remedy for all mischief; no enemy, but a friend; not a cruel tyrant, but a gentle guide, leading us not to mortality, but to immortality, not to sorrow and pain, but to joy and pleasure, and that to endure for ever, if it be thankfully taken and accepted as God's messenger, and patiently borne of us for Christ's love, that suffered most painful death for our love, to redeem us from death eternal."[12]

Moreover, this Anglican sermon also affirmed -

"And we ought to believe, that death, being slain by Christ, cannot keep any man that steadfastly trusteth in Christ, under his perpetual tyranny and subjection: but that he shall rise from death again unto glory at the last day, appointed by Almighty God, like as Christ our head did rise again, according to God's appointment, the third day. . . And to comfort all Christian persons herein, holy Scripture calleth this bodily death a sleep, wherein man's senses be (as it were) taken from him for a season; and yet when he awaketh, he is more fresh than he was when he went to bed. So, although we have our souls separated from our bodies for a season, yet at the general resurrection we shall be more fresh, beautiful, and perfect, than we be now. For now we be mortal, then shall we be immortal; now infected with divers infirmities, then clearly void of all mortal infirmities; now we be subject to all carnal desires, then we shall be all spiritual, desiring nothing but God's glory, and things eternal."[13]

THE IMMEDIATE AFTERLIFE OF THE PILGRIM

John Wesley viewed the souls of mortals, upon death, inheriting *hell* or *hades*, words which the translators of the King James Bible (1611) employed to designate "the invisible world" of the dead.[14] All human souls, upon separating from their bodies, enter into this invisible world. Wesley referred to this place as "the receptacle of separate spirits."[15] Hell, then, claims all souls, whether righteous or not. Moreover, Christ himself went into hell while his body remained in the grave. His soul remained there from his death to his resurrection.[16] However, Wesley argued, hell as the invisible world of the dead is divided into two states: (1) the place of the holy, and (2) the place of the unholy. These states, or places, are kept separated by a great "gulf" so that a soul in one cannot cross over into the other.[17]

The biblical passage behind Wesley's account of hell is Luke XVI, 19-31. The essential parts of the story are as follows - Lazarus (Eleazar in Hebrew), a poor, sick beggar, was laid daily at the gate of a very rich man, probably a pharisee, named Dives by tradition. The beggar received no compassion from the wealthy Dives. As time passed, Lazarus, full of sores, could not keep the dogs from licking his "bare ulcers."[18] Wesley's description adds, ". . .the beggar. . .worn out with hunger and pain and want of all things, died, and was carried by angels into Abraham's bosom."[19] The wealthy man died also and was buried. Wesley supposed that the burial of Dives was attended with great pageantry - ". . .that stupid, senseless pageantry, that shocking insult on a poor, putrefying carcass, was reserved for our enlightened age!"[20] Being in the unholy place of torment, Dives saw Father Abraham afar off in the holy place of the righteous dead with Lazarus being comforted. Dives called across the mighty gulf separating these two places - "Father Abraham, have mercy upon me, and send Lazarus, to dip the tip of his finger in water, and cool my tongue; for I am tormented in this flame!" Wesley's commentary is startling to low-churchmen - "It cannot be denied but here is one precedent in Scripture of praying to departed saints."[21] Abraham's response to Dives was uncompromising - "Son, remember that thou in thy lifetime receivedst thy good things, and likewise Lazarus evil things; but now he is comforted, and thou art tormented. And besides all this, between us and you there is a great gulf fixed." The gulf is forever impassable.[22] While there is more to the Lukan story, this portion is all that is necessary here to mark out Wesley's parameters on the doctrine of hell.

The place of the holy dead, for Wesley, is called both Paradise and Abraham's bosom.[23] Following the *HOMILY AGAINST THE FEAR OF DEATH* - "Abraham's bosom (is) a place of rest, joy, and heavenly consolation"[24] - Wesley argued against the Roman doctrine of Purgatory, in which Christians suffer great pain and torment until they are released into heaven. Wesley's rebuttal is set

forth in his criticism of *THE ROMAN CATECHISM*. His specific
arguments deserve citation -

> ". . .that those that die in a state of grace are yet in a state of torment, and are
> to be purged in the other world, is contrary to Scripture and antiquity... Now
> the elect are justified before they go out of this world; and consequently shall
> have nothing laid to their charge in the next."[25]

> "The state that believers immediately enter upon after death, is said to be
> 'life' for the comfort, and 'everlasting' for the continuance of it. . . As the
> state in which Abraham and Lazarus were, needed no relief; so that in which
> the rich man was, could not obtain it."[26]

> (A quotation from Epiphanius, A.D. 310-403) - "After death is no help to be
> gotten by godliness or repentance. Lazarus doth not there go to the rich man
> nor the rich man unto Lazarus. For the garners are sealed up, and the time is
> fulfilled."[27]

Wesley's sermon on *THE RICH MAN AND LAZARUS*,
preached in Birmingham (March 25, 1788), offers greater interpretation of hell. Paradise or Abraham's bosom is not heaven! It is
"the receptacle of holy souls, from death to the resurrection."[28] No
supposition is more erroneous than that which claims "the souls of
good men, as soon as they are discharged from the body, go directly
to heaven. . . this opinion has not the least foundation in the oracles of God." On the contrary, Wesley argued, Jesus commanded
Mary (St. John XX, 17), after his resurrection -

> "'Touch me not; for I am not yet ascended to my Father' in heaven. He had
> been in Paradise, according to his promise to the penitent thief: 'This day
> shalt thou be with me in Paradise.' Hence it is plain, that Paradise is not
> heaven. It is indeed. . . the ante-chamber of heaven, where the souls of the
> righteous remain till, after the general judgment, they are received into
> glory."[29]

Later in this sermon, Wesley reminded his hearers of an earlier
affirmation - ". . .the word which is here rendered *hell* does not always mean the place of the damned. It is, literally, *the invisible*

world; and is of very wide extent, including the receptacle of separate spirits, whether good or bad."[30] Yet Wesley used the term *hell* in the specialized sense for the place of the damned.[31] As such, it is the place inherited by Dives. So, for Wesley, *hell* is the invisible world for all separated souls, consisting of Paradise or Abraham's bosom for the holy dead, and *hell* for the unholy dead, with a great, impassable gulf between these two states and places. Paradise is also *The Intermediate State*, for Wesley, holding righteous souls from death to the resurrection. Hell for the unholy may also be called *Gehenna*,[32] an Old Testament word meaning "hell-fire." *Life in Abraham's Bosom* - One must prepare for the stay in Paradise. Such a necessity was classically argued by Jeremy Taylor in his *HOLY LIVING AND DYING*, one of Wesley's favorite sources.[33] Certain points of this work appear in Wesley's many treatments concerning the Intermediate State, especially touching one's preparation for entrance. Taylor affirmed -

"He that would die well, must always look for death, every day knocking at the gates of the grave; and then the gates of the grave shall never prevail upon him to do him mischief."[34]

"He that would die well, must, all the days of his life, lay up against the day of death."[35]

"He that desires to die well and happily. . . must be careful that he do not live a soft, delicate, and a voluptuous life; but a life severe, holy, and under the discipline of the cross, under the conduct of prudence and observation, a life of warfare, and sober counsels, labour and watchfulness."[36]

"He that will die well and happily, must dress his soul by a diligent and frequent scrutiny; he must perfectly understand and watch the state of his soul; he must set his house in order, before he be fit to die."[37]

"He that would die well and happily, must, in his lifetime, according to all his capacities, exercise charity; and because religion is the life of the soul, and charity is the life of religion, the same which gives life to the better part of man, which never dies, may obtain of God mercy to the inferior part of man in the day of its dissolution."[38]

In the *EXPLANATORY NOTES UPON THE OLD TESTA-
MENT,* Wesley spoke of the burial plans made by righteous Jacob -
"And he charged them and said unto them, I am to be gathered
unto my people: bury me with my fathers in the cave that is in the
field of Ephron the Hittite" (Genesis XLIX, 29). The commentary
states the expectation of a holy death -

> "I am to be gathered unto my people - Though death separate us from our
> children, and our people in this world, it gathers us to our fathers, and to our
> people in the other world. Perhaps this is why Jacob useth this expression
> concerning death, as a reason why his sons should bury him in Canaan, for
> (saith he) I am to be gathered unto my people, my soul must be gone to the
> spirits of just men made perfect, and therefore bury me with my fathers
> Abraham and Isaac, and their wives."[39]

When the souls of holy persons are ready to depart their dying
bodies, Wesley believed, the angels of God come to escort each
soul in a convoy to Paradise or Abraham's bosom. Even Lazarus,
the poor beggar, was given this angelic transport. In some instances, Wesley acknowledged, angelic music has been heard by
those attending the death of a saint.[40] For the good man entering
eternity, "the convoy attends, the ministering host of invisible
friends."[41] The angelic friends receive "the new-born spirit," and
they conduct him safe "into Abraham's bosom," to enjoy "the de-
lights of Paradise; the garden of God." There in this spiritual Eden
the light of God's countenance "perpetually shines." This holy
place is, in Wesley's terminology, "the ante-chamber of heaven"[42]
and "the porch of heaven."[43]

The souls of the righteous in Abraham's bosom are immediately
blessed upon their arrival. "There the wicked cease from troubling,
there the weary are at rest." These blessed souls have "numberless
sources of happiness which they could not have upon earth."
Foremost of these is meeting "the glorious dead of ancient days."
They are able to converse with Adam, Noah, Abraham, Moses, the
Prophets, the Apostles, and the saints of all ages.[44] Such communion in Paradise serves as a continual "ripening for heaven."[45]

Furthermore, the communion of saints in Paradise means that personal identity is not lost, as some have boldly declared. If there are no identities, Wesley argued, how could there be Adam, Noah, Abraham, Moses, Prophets, or Apostles to meet and engage in conversation? Dives recognized Abraham and Lazarus, although a great gulf separated him from them. Wesley maintained that believers do recognize one another in Abraham's bosom, and that family members enjoy happy reunions. Moreover, he held, souls in Paradise frequently visit loved ones still on earth as part of the reality called The Communion of the Saints -

"It has in all ages been allowed that the communion of the saints extends to those in Paradise, as well as those on earth, as they are all one body united under one Head, and

    'Can death's interposing tide Spirits one in Christ divide? But it is difficult to say, either what kind, or what degree of union, may be between them. It is not improbable, their fellowship with us is far more sensible than ours with them. Suppose any of them are present, they are hid from our eyes, but we are not hid from their sight. They, no doubt, clearly discern all our words and actions, if not all our thoughts too. For it is hard to think these walls of flesh and blood can intercept the view of an angel being. But we have, in general, only a faint and indistinct perception of their presence, unless in some peculiar instances, where it may answer some gracious ends of Divine Providence. Then it may please God to permit, that they should be perceptible, either by some of our outward senses, or by an internal sense, for which human language has not a name. But I suppose this is not a common blessing...I have known but few instances of it. To keep up constant and close communion with God is the most likely means to obtain this also."[46]

One additional passage from Wesley at this point is pertinent -

"I have heard my mother say, 'I have frequently been as fully assured that my father's spirit was with me, as if I had seen him with my eyes.' But she did not explain herself any further. I have myself many times found on a sudden so lively an apprehension of a deceased friend, that I sometimes turned about to look: At the same time I have felt an uncommon affection for them. But I never had anything of this kind with regard to any but those that died in the faith. In dreams, I have had exceeding lively conversations with them; and I doubt not but they were then very near."[47]

Why should souls in Abraham's bosom make interceptions with persons living on earth? Wesley's answer was given as part of his sermon *ON FAITH.* The souls of the just, "generally lodged in Paradise. . .may. . . in conjunction with angels," minister to those who are heirs of salvation. Wesley quoted an appropriate verse upon this subject - "May they not

> Sometimes, on errands of love,
> Revisit their brethren below?

The holy souls in Abraham's bosom enjoy the greatest communion, with angels and "the wisest and best men that have lived from the beginning of the world."[48] More important in this communion than angels, Abraham, or Moses, is Christ himself. It is he who gives these souls their rest.[49] They learn more from him in one hour than they learned during a lifetime on earth.[50] It is his purpose to make them perfect in preparation for heaven.[51] Under this tutorage, Wesley claimed, "human spirits swiftly increase in knowledge, in holiness, and in happiness; conversing with all the wise and holy souls that lived in all ages and nations from the beginning of the world." Their conversation extends to angels and archangels, and they freely speak "with the eternal Son of God, 'in whom are hid all the treasures of wisdom and knowledge.'"[52] Whatever these good souls learn, they retain forever. They forget nothing.[53] They joyously await the resurrection and their glorification in the immediate presence of God.

Instances of Holy Dying - The writings of John Wesley abound with stories of saints victorious in death, taking the step from mortality to immortality in Abraham's bosom. The *JOURNAL* especially treats this subject, and the years covered by volumes III through VI possesses the greatest number of these stories.

Upon her death bed in Bristol, Molly Thomas was overwhelmed with the pains of body and soul. But on June 6, 1745, Christ suddenly spoke healing peace to her heart. Wesley recorded this event -

"In that hour she cried out, 'Christ is mine! I know my sins are forgiven me.' Then she sung praise to Him that loved her, and bought her with His own blood. The fear of death was gone, and she longed to leave her father, her mother, and all her friends. She said, 'I am almost at the top of the ladder; now I see the towers before me, and a large company coming up behind me: I shall soon go. Tis but for Christ to speak the words, and I am gone; I only wait for that word, Rise up, my love, and come away. When they thought her strength was gone, she broke out again:

> Christ hath the foundation laid,
> And Christ shall build me up:
> Surely I shall soon be made
> Partaker of my hope.
>
> Author of my faith He is;
> He its finisher shall be:
> Perfect love shall seal me His
> To all eternity.

So she fell asleep."[54]

Another incident of triumph, dated October 31, 1766 in the *JOURNAL*, describes Wesley's hasty flight to Leytonstone to arrive before the death of a particular saint -

"Some hours before she witnessed that good confession -

> Nature's agony is o'er
> And cruel sin subsists no more.

Awhile after she cried out earnestly, 'Do you not see Him? There He is! Glory! glory! glory! I shall be with Him forever - for ever - for ever! So died Margaret Lewen! A pattern to all young women of fortune in England: a real Bible Christian. So she 'rests from her labours, and her works do follow her.'"[55]

On February 10, 1769, Wesley went to Deptford to visit William Brown, who was slowly dying - "worn out with age and pain, and long confined to his bed, without the use of either hand or foot."

But the scene was not one of tragedy because Wesley found him to have "the use of his understanding and his tongue, and (he) testifies that God does all things well." Moreover, Wesley articulated, "He has no doubt or fear, but is cheerfully waiting till his change shall come."[56]

Children may die as triumphantly as adults, being met by the angels and escorted into Paradise. One of Wesley's books for children was the *TOKENS FOR CHILDREN*. It consisted of ten carefully abridged stories of spiritual awakening, piety, and dying of young children, originally having been written by James Janeway.[57] It is interesting to note that Wesley withdrew the *TOKENS* from circulation among the Methodists shortly after the Annual Conference of 1744. The death theme in the *TOKENS* proved to be much too melancholy although they were intended to amplify a triumphant tone. Nevertheless, Wesley slipped such stories into his *JOURNAL* whenever he could, but without lengthy descriptions. For example -

"Sunday, February 1771 - I buried the remains of Joan Turner, who spent all her last hours in rejoicing and praising God, and died full of faith and of the Holy Ghost, at three and a half old."[58]

One of Wesley's favorite death stories involved the passing of Charles Greenwood. Besides being printed in the *JOURNAL*, which version appears here, it appeared in longer form in the *ARMINIAN MAGAZINE* (1783). Moreover, Charles Wesley wrote a version of this event.[59] The remarkable feature in the account is the overcoming of the fear of death -

"To-day Charles Greenwood went to rest. He had been a melancholy man all his days, full of doubts and fears, and continually writing bitter things against himself. When he was first taken ill he said he should die, and was miserable through fear of death; but two days before he died the clouds dispersed, and he was unspeakably happy, telling his friends, 'God has revealed to me things which it is impossible for man to utter.' Just when he died, such glory filled the room that it seemed to be a little heaven; none could grieve or shed a tear, but all present appeared to be partakers of his joy."[60]

The fear of death frequently stalks the pilgrim, although Abraham's bosom is his destination, and despite the angelic convoy.

The Fear of Death and the Pilgrim - The fact that Methodist pilgrims were intellectually prepared by Wesley for death did not provide an easy escape from the fear of death. In many cases, Wesley's sermons, treatises, and commentaries on Abraham's bosom were well known and believed, but when death approached more than one Wesleyan pilgrim became "horribly afraid" and fell into what Wesley called "the lowest darkness and in the deep."[61] Wesley's response was to preach more upon the theme of Paradise, as his *JOURNAL* clearly shows, and to counsel his suffering followers. He also abridged the writings of others upon the subject, including them in the *CHRISTIAN LIBRARY* and *ARMINIAN MAGAZINE*. While these efforts were no doubt successful, he maintained that God often imparts a special revelation of Paradise to the Christian pilgrim and thus Himself dispels fear prior to death.[62]

The sermon *ON THE FEAR OF DEATH*, found in the *HOMILIES OF THE CHURCH OF ENGLAND*, cited three causes why "worldly men" fear death: (1) because "they shall lose thereby their worldly honours, riches, possessions, and all their hearts desires;" (2) because "of the painful diseases, and bitter pangs, which common men suffer, either before or at the time of death;" and (3) the chief cause, which is "the dread of the miserable state of eternal damnation, both of body and soul, which, they fear shall follow, after their departing from the bodily pleasures of this present life." However, there is a sharp distinction to be made between the "worldly man" and the Christian pilgrim. The sermon gives an eloquent description of the true Christian -

"There is never a one of all these causes, no, nor yet them altogether, that can make a true christian man afraid to die; but plainly contrary, he conceiveth great and many causes, undoubtedly grounded upon the infallible and everlasting truth of the word of God, which moveth him not only to put away the fear of bodily death, but also, for the manifold benefits and singular commodities, which ensue unto every faithful person by reason of the same, to wish, desire, and long heartily for it. For death shall be to him no death at all, but a very deliverance from death, from all pains, cares, and sorrows, mis-

eries, and wretchedness of this world, and the very entry into rest, and a beginning of everlasting joy, a tasting of heavenly pleasures, so great, that neither tongue is able to express, neither eye to see, nor ear to hear them; no,
nor any earthly man's heart to conceive them."[63]

From his earliest years Wesley knew this sermon. It was
preached by his father Samuel in St. Andrews Church in Epworth,
as were the other sermons of the *HOMILIES*. However, when he
was aboard ship, on the high seas from November 1735 onward,
going to colonial Georgia in the New World, great storms lashed
the ship, washing over the entire deck and into cabins and holds.
One night, as he related it, "I was awaked by the tossing and roaring
of the wind, and plainly showed I was unfit, for I was unwilling, to
die."[64] Two months later, still in passage to Georgia, another storm
struck Wesley's ship -

"It rose higher and higher until nine (p.m.). About nine the sea broke over us
from stem to stern; burst through the windows of the state cabin (Mr.
Oglethorpe's cabin, where three or four of us were sitting with a sick woman,
and covered her all over.) A bureau sheltered me from the main shock...
About eleven I lay down in the great cabin, and in a short time fell asleep,
though very uncertain whether I should awake alive, and much ashamed of
my unwillingness to die. O how pure in heart must he be who would rejoice
to appear before God at a moment's warning! Toward morning 'He rebuked
the winds and the sea, and there was a great calm.'"[65]

The next day, Sunday, January 18, 1736, Wesley conducted a
shipboard service of thanksgiving to God for his deliverance from
the storm. Only a few persons participated, the others, called
"cowards" by Wesley, "denied we had been in any danger." He resolved never to trust "in the sovereign, saving grace of fear."[66] He
had hoped that the fear of death, which the other passengers and
crew members had experienced, would bring them to faith in God,
following the deliverance of the ship. It did not. The next Friday
evening, another storm began. It was extremely violent. In the
morning the crew had no choice but to let the ship drive herself

through wind and wave. Wesley's account is graphic in describing the terror of the sea -

"I could not but say to myself, 'How is it that thou hast no faith?' being still unwilling to die. About one in the afternoon, almost as soon as I had stepped out of the great cabin door, the sea did not break as usual, but came with a full, smooth tide over the side of the ship. I was vaulted over with water in a moment, and so stunned that I scarce expected to lift up my head again till the sea should give up her dead. But, thanks be to God, I received no hurt at all. About midnight the storm ceased."[67]

One thing often overlooked by some persons who interpret Wesley as a gross sinner at this stage of life, greatly afraid to die, is the spirit of thanksgiving which seized him after being delivered from these storms. If willingness to die is the mark of the true Christian, why should pilgrims pray to be delivered from terrifying wind and wave? And upon being delivered, why should they give thanks? Put these judgmental interpreters of Wesley aboard a small sloop in the North Atlantic from November to January - without stabilizers on the hull, electric pumps to keep the bilge from filling, a radio to call for help, and modern flotation devices - then behold their unwillingness to die, and observe their thanksgiving should they survive the flow of violent storms.

A fourth great storm overtook the Simmonds, Wesley's ship commanded by Captain Joseph Cornish.[68] This one came just two days following the third storm. The *JOURNAL* reads -

"While the calm continued I endeavoured to prepare myself for another storm. At noon (another) storm began. At four it was more violent than any we had had before. Now, indeed, we could say, 'The waves of the sea were mighty, and raged horribly. They rose up to the heavens above, and clave down to hell beneath.' The winds roared round about us, and - what I never heard before - whistled as distinctly as if it had been a human voice. The ship not only rocked to and fro with the utmost violence, but shook and jarred with so unequal, grating a motion, that one could not but with great difficulty keep one's hold of anything, nor stand a moment without it. Every ten minutes came a shock against the stern or side of the ship, which one would think should dash the planks in a thousand pieces... We spent two or three hours after prayers in conversing suitably to the occasion, confirming one another in a calm submission to the wise, holy, gracious will of God. And now a storm did not appear so terrible as before. Blessed be the God of all consolation, who alone doeth wonders, and is able mightily to deliver His people."[69]

Prayer and confirming one another "in a calm submission to the wise, holy, gracious will of God," brought Wesley out of his fear of death. The Moravians aboard the Simmonds, fellow pilgrims to the New World, had consistently braved these storms without fear or anxiety as the *HOMILY* extolled. But, alas, for Wesley, he had to subdue this dreadful fear every time a storm arose - which he did. After visiting these German pietists and observing their trust in God in the face of death, Wesley went to his fellow countrymen aboard ship and spoke boldly of the difference between those who fear God and those who do not fear God. His entry in the JOURNAL for this day reads, "This was the most glorious day which I have hitherto seen."[70] In his Savannah Journal, Wesley recounted one of these storms - "I was at first afraid; but cried to God, and was strengthened. Before ten I lay down; I bless God without fear.[71] A victory had been achieved, but the fear of death was not yet completely conquered.

Once in Savannah, looking back over his motives and labors, Wesley spoke of the fear that death posed from the outset of coming to Georgia. One hundred and sixty leagues (480 nautical miles) from Land's End (England), he was quite introspective and wrote dejectedly -

"I have a fair summer religion. I can talk well; nay, and believe myself, while no danger is near. But let death look me in the face, and my spirit is troubled. Nor can I say, 'To die is gain' -

I have a sin of fear, that when I've spun
My last thread, I shall perish on the shore!

. . .Oh, who will deliver me from this fear of death? What shall I do? Where shall I fly from it? Should I fight against it by thinking or by not thinking of it? A wise man advised me some time since, 'Be still and go on.' Perhaps this is best, to look upon it as my cross; when it comes, to let it humble me, and quicken all my good resolutions, especially that of praying without ceasing; and at other times, to take no thought about it, but quietly to go on in the work of the Lord."[72]

Two years and nearly four months later, after having faced many dangers intermittently with fear and trust, Wesley concluded himself a sinner, under the sentence of death, a child of wrath, and an heir of hell.[73] He was certain now that he had reason to fear death. Death would mean the punishment of Gehenna rather than the comfort of Paradise.

Upon returning to England, Wesley spent considerable time reviewing the events of the Georgia expedition. God, he believed, "hath in some measure humbled me and proved me, and shown me what was in my heart. . . hereby I am delivered from the fear of the sea, which I had both dreaded and abhorred from my youth."[74] The fear of death had persisted and attacked him on the voyage back to England. He reasoned that this vacillation between fear and faith was the result of latent unbelief, and "that the gaining a true, living faith was the 'one thing needful' for me."[75] Some students of Wesley maintain that the Aldersgate experience ended the reign of the fear of death and the sea, as a benefit of saving faith. No doubt that was what Mr. Wesley believed. On September 13, 1743, Wesley left by a small fishing vessel for the Isles of Scilly. At long last, Wesley was completely dead to the fear of sea and death -

"It seemed strange to me to attempt going, in a fisher boat, fifteen leagues upon the main ocean; especially when the waves began to swell, and hang over our heads. But I called to my companions, and we all joined together in singing lustily and with a good courage:

> When passing through the wat'ry deep,
> I ask in faith His promised aid;
> The waves an awful distance keep,
> And shrink from my devoted head;
> Fearless their violence I dare:
> They cannot harm - for God is here."[76]

The origin of this song is not known. It has been suggested that Wesley wrote it. Whatever the case may be, at this point, Wesley and his sailing companions knew it by heart and sang it in the midst of high seas. Had Wesley become a true Christian without fear of death, or, as some cynics have asked, was he now a better sailor?

In November of 1753, Wesley was suffering from a respiratory ailment and was rapidly wasting away. His physician, Dr. Fothergill, advised he get into the country immediately for the purer air, for rest, for a special diet of asses' milk, and for daily horseback riding to stimulate the flow of blood through the lungs. Wesley's JOURNAL gives the account of this flight -

> "So (not being able to sit a horse) about noon I took coach for Lewisham. In the evening (not knowing how it might please God to dispose of me), to prevent vile panegyric, I wrote as follows:
>
> Here Lies The Body
>
> of
>
> JOHN WESLEY
>
> A BRAND PLUCKED OUT OF THE BURNING:
>
> WHO DIED OF A CONSUMPTION IN THE FIFTY-FIRST YEAR OF HIS AGE,
> NOT LEAVING, AFTER HIS DEBTS ARE PAID,
> TEN POUNDS BEHIND HIM:
>
> PRAYING,
>
> GOD BE MERCIFUL TO ME, AN UNPROFITABLE SERVANT!
>
> Wed. 28. - I found no change for the better, the medicines which had helped me before now taking no effect. About noon (the time that some of our brethren in London had set apart for joining in prayer) a thought came into my mind to make an experiment. So I ordered some stone brimstone to be powdered, mixed with the white of an egg, and spread on brown paper, which I applied to my side. The pain ceased in five minutes, the fever in half an

hour, and from this hour I began to recover strength. The next day I was able to ride, which I continued to do every day till January 1."[77]

Although Wesley wanted to live, and did everything medically possible to live, he was not afraid of death. His epitaph was written exactly as he wished it to appear on his white stone - a witness to his trust in a merciful God. However, the epitaph was premature by thirty-eight years! And, he frequently cited St. Paul - "For me to live is Christ, and to die is gain!" When he, at eighty-eight years of age, finally lay upon his deathbed, his departing words were,

"The best of all is, God is with us. . . The best of all is, God is with us.. It will not do, we must take the consequence: never mind the poor carcass. . . Who are these? (his sight, now almost gone, preventing him from knowing his most intimate friends. . .except by their voices). . . It is the Lord's doing, and marvelous in our eyes. . . He giveth his servants rest. . . We thank Thee, O Lord, for these and all Thy mercies; bless the Church and King: grant us truth and peace, through Jesus Christ our Lord, for ever and ever. . . He causeth His servants to lie down in peace. . . Amen. . . The clouds drop fatness. . . The Lord is with us, the God of Jacob is our refuge. . . I'll praise - I'll praise - I' . . . Farewell!"[78]

Elizabeth Ritchie, an eyewitness to Wesley's "change" from this life to Abraham's bosom, added the following to her written account -

"We felt what is inexpressible; the ineffable sweetness that filled our hearts as our beloved Pastor, Father, and Friend entered his Master's joy, for a few moments blunted the edge of our painful feelings on this truly glorious, melancholy occasion. As our dear aged Father breathed his last, Mr. Bradford was inwardly saying, 'Lift up your heads, O ye gates; be ye lift up, ye everlasting doors, and let this heir of glory enter in.' Mr. Rogers gave out:

'Waiting to receive our spirit,
Lo. the Saviour stands above;
Shows the purchase of His merit,
Reaches out the crown of love.'

I then said, 'Let us pray for the mantle of our Elijah;' on which Mr. Rogers prayed in the spirit for the descent of the Holy Ghost on us, and all who mourn the general loss the Church Militant sustains by the removal of our much-loved Father to his great reward. Even so. Amen."[79]

So, John Wesley the pilgrim passed through the door of death, without fear, entering into the paradisiacal bosom of Abraham.

THE IMMEDIATE AFTERLIFE OF THE NONPILGRIM

"Where their worm dieth not, and the fire is not quenched."
Mark IX, 48.

The above verse was the text for Wesley's sermon *OF HELL*. He composed the sermon during his pre-Georgia experience. Nehemiah Curnock was correct when he stated that the young Wesley was very much a scholar when he wrote this sermon, treating the theme of hell in its broadest biblical context, supported with sound logic.[80] Some scholars have argued that the sermon was later discarded by Wesley because he came to believe that a merciful God and hell are incompatible. They can cite no explicit passage from Wesley in support of this contention. In fact, they overlook numerous occasions between 1764 and 1781 when Wesley preached this very sermon.[81] A survey of the sermon will prove beneficial to this aspect of our study.

1. The text is repeated by Jesus three times between verses 43 and 48, after "If thy hand offend thee," "If thy foot affend thee," and "If thy eye..." Jesus, by using repetition here, meant to show the importance of avoiding the punishment of hell in the next life.[82]

2. These warnings of Jesus were meant for all men, for "enormous sinners" and for "the holiest men upon the earth."[83] "I say unto you, my friend, Fear not them that can kill the body, and after that have no more that they can do. But I say unto you, Fear him, who after he hath killed hath power to cast into hell." "...fear lest he should cast you into the place of torment. And this very

fear, even in the children of God, is one excellent means of pre-
serving them from it."[84]

3. Christians should know the "oracles of God" concerning the
"future state of punishment." In fact, the scriptural account is far
superior to the heathen and childish account given the future state
by Virgil and Juvenal who lamented, "Even our children do not believe a word of the tales concerning another world."[85]

4. The biblical doctrine of hell is "worthy of God, the Creator,
the Governor of mankind. . . suitable to His wisdom and justice by
whom 'Tophet was ordained of old.'" Tophet, for Wesley, was another name for Gehenna. It was "originally prepared, not for the
children of men, but 'for the devil and his angels.'" Wesley wondered greatly why so many mortals should shun all warnings about
hell and "resolve to have their portion with the devil and his angels." These persons suffer as "either *paena damni*, - 'what they
lose;' or *paena sensus*, - 'what they feel.'"[86]

a. *paena damni* - Wesley argued from classical theology that
those who suffer this condition experience "the punishment of
loss." They begin this loss as soon as their souls leave their bodies.
The soul loses the pleasures associated with the bodily senses -
"The smell, the taste, the touch, delight no more: The organs that
ministered to them are spoiled, and the objects that used to gratify
them are removed far away." In Gehenna, "the dreary regions of
the dead," all these pleasures are forgotten, or if remembered,
"remembered with pain." Imaginary pleasures are also abolished
forever. "There is no grandeur in the infernal regions; there is
nothing beautiful in those dark abodes; no light but that of livid
flames." There is nothing new in Tophet, "but one unvaried scene
of horror upon horror!" The only music there is "that of groans and
shrieks of weeping, and gnashing of teeth; of curses and blasphemies against God, or cutting reproaches of one another." A
sense of honor does not exist there, and all Gehenna's residents are
"heirs of shame and everlasting contempt." Moreover, these souls
are "totally separated from all the things they were fond of in the
present world." This is especially true of friends and loved ones -
"They are torn away from their nearest and dearest relations; their

wives, husbands, children; and. . . the friend which was as their own soul." The pleasures of love and friendship are forever lost and "vanished away." There "is no friendship in hell." But, they shall suffer a greater loss - "their place in Abraham's bosom." It never before entered their minds "what holy souls enjoy in the garden of God, in the society of angels, and of the wisest and best men that have lived from the beginning of the world."[87] Only then will they "fully understand the value of what they have vilely cast away."

Meanwhile, the souls of Abraham's bosom are preparing for a far greater happiness. "For paradise is only the porch of heaven; and it is there the spirits of just men are made perfect. It is in heaven only that there is fullness of joy; the pleasures that are at God's right hand for evermore." Consequently, the loss of Abraham's bosom and then heaven, by unholy souls, "will be the completion of their misery." Following the argument of St. Augustine, Wesley further articulated, "They will then know and feel, that God alone is the centre of all created spirits; and, consequently, that a spirit made for God can have no rest out of him." Banishment "from the presence of the Lord is the very essence of destruction to a spirit that was made for God... And if that banishment lasts forever, it is 'everlasting destruction.'" If this were not enough punishment, "to the punishment of loss, will be added the punishment of sense."

b. *paena sensus* - Whatever unholy souls experience by the punishment of loss, that is inferior to what they feel. The history of burial practices, Wesley noted, indicates an attempt to return the body of man back to "the general mother, earth." Dust to dust was the universal practice, placing the body in the ground and covering it with earth. Later, "another method obtained, chiefly among the rich and great, of burning the bodies... and frequently in a grand magnificent manner... they erected huge funeral piles, with immense labour and expense." By these methods the body of man was reduced to "its parent dust." In either case, "the worm or the fire soon consumed the well-wrought frame; after which the worm itself quickly died, and the fire was entirely quenched." But, Wesley continued, "there is, likewise, a worm that belongeth to the future state; and that is a worm that never dieth! and there is a fire hotter than that of the funeral pile; and it is a fire that will never be quenched!"[88]

The "worm that never dieth," Wesley argued, "seems to be a guilty conscience." As such, it includes "self-condemnation, sorrow, shame, remorse, and a sense of the wrath of God." In addition to these, the condemned soul feels all the unholy passions - "fear, horror, rage, evil desires; desires that can never be satisfied." Add also unholy tempers - "envy, jealousy, malice, and revenge." These incessantly "gnaw" on the soul, "as the vulture was supposed to do the liver of Tityus." Should one add to all of these the "hatred of God, and all his creatures; all these united together may serve to give us some little, imperfect idea of the worm that never dieth."

Moreover, Jesus spoke of two parts of the future punishment - "'Where their worm dieth not,' of the one; 'where the fire is not quenched,' of the other." What is the difference?

"Does it not seem to be this? *The fire* will be the same, essentially the same, to all that are tormented therein; only perhaps more intense to some than others, according to their degree of guilt: but *their worm* will not, cannot be the same; It will be infinitely varied, according to their various kinds, as well as degrees, of wickedness. This variety will arise partly from the just judgment of God, 'rewarding every man according to his works.' For we cannot doubt but this rule will take place no less in hell than in heaven. As in heaven 'every man shall receive his own reward,' . . . so undoubtedly, every man, in fact, will receive his own bad reward, according to his own bad labour. . . . Variety of punishment will likewise arise from the very nature of the thing. As they that bring most holiness to heaven will find most happiness there; so, on the other hand, it is not only true, that the more wickedness a man brings to hell the more misery he will find there; but that this misery will be infinitely varied according to the various kinds of his wickedness. It was therefore proper to say, the fire, in general; but *their worm*, in particular."[89]

Is there material fire in hell? Wesley's logic followed the deductive method of Aristotle - ". . .if there be any fire, it is unquestionably material. For what is immaterial fire? The same as immaterial water or earth! Both the one and the other is absolute nonsense, a contradiction in terms." The deduction continues, "Either, there-

fore, we must affirm it to be material, or we deny its existence. But
if we granted them, there is no fire at all there, what would they
gain thereby? seeing this is allowed, on all hands, that it is either
fire or something worse." Wesley raised some questions of his own
- "Does not our Lord speak as if it were real fire?" "Is it possible
then to suppose that the God of truth would speak in this manner,
if it were not so?" "Does he design to fright his poor creatures?"
"What, with scarecrows?" "With vain shadows of things that have
no being?" Wesley then exhorted - "O let not any one think so!
Impute not such folly to the Most High!"[90]

Another popular objection to the hell-fire was thrown into the
sermon and refuted by Wesley. "It is not possible that fire should
burn always. For by the immutable law of nature, it consumes
whatever is thrown into it. . . as soon as it has consumed its fuel, it
is itself consumed; it goes out." Wesley's response represents a
dynamic synthesis of biblicism and logic -

> "It is most true, that in the present constitution of things, during the present
> laws of nature, the element of fire does dissolve and consume whatever is
> thrown into it. But here is the mistake: The present laws of nature are not
> immutable. When the heavens and the earth shall flee away, the present
> scene will be totally changed; and, with the present constitution of things, the
> present laws of nature will cease. After this great change, nothing will be dis-
> solved, nothing will be consumed any more. Therefore, if it were true that
> fire consumes all things now, it would not follow that it would do the same af-
> ter the whole frame of nature has undergone that vast, universal change."[91]

Furthermore, Wesley argued, fire does not consume *all* things
now. Aristotle had taught his students to find one exception to any
premise that uses the modifier "all." One exception calls in doubt
the universality of the premise, rendering it questionable. Wesley
cited the existence of *Linum Asbestum*, the incombustible flax re-
cently added to the collection of the British Museum - "If you take
a towel or handkerchief made of this, you may throw it into the
hottest fire, and when it is taken out again, it will be observed,
upon the nicest experiment, not to have lost one grain of its
weight." The logical conclusion, therefore, is - "Here, therefore, is

a substance before our eyes, which, even in the present constitution of things, may remain in fire without being consumed."[92] Could the soul be similar to asbestos in the future punishment, suffering in the flames but without being consumed?

Wesley totally rejected the extra-biblical exaggerations of the sufferings in hell which were popular in his day. For instance, he criticized Thomas a'Kempis for teaching that misers have melted gold poured down their throats in hell. "This is too awful a subject to admit of such play of imagination. Let us keep to the written word. It is torment enough to dwell with everlasting burnings."[93]

5. The inhabitants of hell are perfectly wicked, "having no spark of goodness remaining . . .and they are restrained by none from exerting to the uttermost their total wickedness." Fellow prisoners in hell cannot restrain another, nor themselves. Nor will God restrain them from wickedness, seeing he has forgotten them and delivered them "over to the tormentors."[94] Their angelic tormentors "vary their torments a thousand ways."[95]

6. The torments of body and soul of the unholy "are without intermission. They have no respite from pain; but 'the smoke of their torment ascendeth up day and night.'" Day and night in Gehenna, however, are as one continuous night of darkness, having "no interruption of their pain. No sleep accompanies that darkness." Despite Home and Milton, there is no sleep in either hell or heaven. That being the case, there is no possibility of souls in hell "fainting away" because of the great torments they suffer there.[96] While the inhabitants of earth are frequently diverted from their sufferings, "the inhabitants of hell have nothing to divert them from their torments, even for a moment." There is no change of seasons, companions, or business. The duration of hell has no end. "Nothing but eternity is the term of their torment." Wesley commented on the duration of eternity in concluding this sermon - "Suppose millions of days, of years, of ages elapsed, still we are only on the threshold of eternity! Neither the pain of body or of soul is any nearer an end, than it was millions of ages ago."[97] Wesley's final appeal to his congregation was a strong exhortation -

"O let us look back and shudder at the thoughts of that dreadful precipice, on the edge of which we have so long wandered! Let us fly for refuge to the hope that is set before us, and give a thousand thanks to the divine mercy, that we are not plunged into this perdition!"[98]

In his correspondence with William Law, dated January 6, 1756, John Wesley recapitulated some of the arguments of the sermon analyzed above. But, he recited a list of proof-texts from the Bible that gave foundation to his arguments. Following which he applied this argument -

"Now, thus much cannot be denied, that these texts speak as if there were really such a place as hell, as if there were a real fire there, and if it would remain for ever. I would then ask but one plain question: If the case is not so, why did God speak as if it was? Say you, 'To affright men from sin?' What, by guile, by dissimulation, by hanging out false colours? Can you possibly ascribe this to the God of truth? Can you believe it of Him? Can you conceive the Most High dressing up a scarecrow, as we do to fright children? Far be it from him! If there be then any such fraud in the Bible, the Bible is not of God. And indeed this must be the result of all: If there be 'no unquenchable fire, no everlasting burnings,' there is no dependence on those writings wherein they are so expressly asserted, nor of the eternity of heaven, any more than of hell. No hell, no heaven, no revelation!"[99]

The contrast between the death of the righteous and the unrighteous is great. Following Luke 16, Wesley believed the good angels of God come to escort the soul into Abraham's bosom. Conversely, he argued from reason, at the moment of separation, when the soul leaves the body, evil angels come to seize the unholy soul and drag it into hell for its uninterrupted punishment and torment.[100] But, he argued, "this is not the nethermost hell." It is a holding place until the great judgment, when it shall be cast into the "the lake of fire" which burns for ever and ever. The "lake" is Wesley's "nethermost hell." Gehenna, therefore, is only temporary. The unholy souls in Gehenna, however, "carry with them their own hell, in the worm that never dieth." Wesley added one more probability to the activities of unholy souls after death - "Till then (day of judgment) they will probably be employed by their bad master in advancing his infernal kingdom, and in doing all the mischief that lies in their power, to the poor, feeble children of men."[101] By this Wesley meant that many evil souls after death are not locked into hell, but they are made ghosts to do Satan's work of mischief. Evil ghosts, for Wesley, are a reality, as are good ghosts.

GHOSTS AND APPARITIONS

"How different, alas! is the case with him who loses his own soul! The moment he steps into eternity, he meets with the devil and his angels. Sad convoy into the world of spirits! Sad earnest of what is to come! And either he is bound with chains of darkness, and reserved unto the judgment of the great day; or, at best, *he wanders up and down*, seeking rest, but finding none." (Evil ghosts) ". . .See where the servants of the Lord, a busy multitude appear; For Jesus day and night employed, His heritage they toil to clear." (Good ghosts).[102]

It has been a common practice in American Methodism of late to treat Wesley's belief in ghosts as a matter of great superstition. Some interpreters try, however, to soften the charge somewhat by claiming Wesley to have been a child of his age - "from our stance in more modern times we now know that such a belief is superstition and nothing more." This is the explicit position taken by Bishop Francis J. McConnell in his influential book on Wesley.[103] But, as McConnell rightly observed, Wesley held his belief in ghosts as having its genesis in the Bible. It may be observed still further that he added to the biblical witness to ghosts that of the Fathers of the early Church, reason, and universal Christian experience. Wesley's understanding of this phenomenon was theological in nature, rather than superstitious. He viewed the subject from philosophical perspectives as well, reasoning from an Aristotelian interplay of physics and metaphysics. It is simply intellectual sloth to accept Wesley's treatment of ghosts and apparitions as sheer superstition. The greatest superstition may well be the naive acceptance of modernity's rejection of any metaphysical influence on the physical. What will history say in evaluation of the Twentieth Century and its commitment to materialism - a century marked by intellectual rejection of metaphysics, marked by world wars and violent revolutions, marked by bloodletting in both holocaust and abortion clinic, marked by universal binges of consumerism and hedonistic pleasure - What will history say in judgment of us? Perhaps it will be more tolerable for Sodom and Gomorrah in the day of Yahweh than for this present age!

John Wesley's starting point for his understanding of ghosts and apparitions was clearly the Bible. The foremost passage, for ghosts, was I Samuel XXVIII. King Saul visited the witch of Endor in order to interview the departed spirit of Samuel. Wesley's observation on this purpose reads, "This practice of divination by the dead, or the souls of dead persons, was very usual among all nations."[104] As the seance progressed, the witch was startled to see "Samuel, instead of the spirit whom she expected to see." Here Wesley explained the ghostly appearance of Samuel as a divinely ordained act of God to advance his glory.[105] The witch exclaimed, "I saw gods ascending out of the earth." Wesley's comment of "gods" is pertinent here - "That is, a god, and divine person, glorious, and full of majesty and splendor, exceeding not only mortal men, but common ghosts."[106] Samuel, by virtue of his holy life, was in death a "god" to be sent briefly by God back to the living for providential purposes. "Common ghosts" were, for Wesley, the restless souls of departed unbelievers. But ghosts like Samuel temporarily leave their rest in paradise to bring some divine benefit to mortals still in the earthly life.

The phrase "ascending out of the earth" suggested to Wesley another important Old Testament passage - Job XXVI, vss. 5 and 6. Job's response to Bildad contained what Wesley believed to be a description of hell from whence evil or common ghosts arise. The Hebrew word *sheol* in the context of this passage means "the nether world" and implies the "shades (departed souls). . .huddled together."[107] Wesley's commentary on this section recognizes that God has dominion even in sheol -

"Dead things. . .or. . .dead men, and of the worst of them, such as died in their sins, and after death were condemned to further miseries; for of such this very word seems to be used. Prov. II, 18, IX, 18. who are here said to mourn or groan from under the waters; from the lower parts of the earth, or from under those subterranean waters, which are supposed to be within and under the earth; Psalm XXXIII, 7. and from under the inhabitants thereof; either of the waters or of the earth, under which these waters are, or with the other inhabitants thereof; of that place under the waters, namely, the apostate spirits. So the sense is, that God's dominion is over all men, yea, and even the dead, and the worst of them, who though they would not own God, nor his providence, while they lived, yet now are forced to acknowledge and feel that power which they despised, and bitterly mourn under the sad effects of it in their infernal habitations. ...Hell - Is in his presence, and under his providence. Hell itself, that place of utter darkness, is not hid from his sight."[108]

Before proceeding further on the subject of ghosts, a definitive statement concerning the nature of apparitions is sorely needed. This is the case since Wesley constantly linked the two. An apparition is an appearance by God, an angel, or a departed saint to a person on earth. Such appearances are sensible and differ from visions and revelations which are intellectual in nature and content. The person who experiences an apparition is convinced that there is a real contact, with an actual dialogue of sorts between the other-worldly visitor and the experiencer. The biblical record of apparitions is lengthy. Wesley, as noted earlier, started with Christ appearing to Adam and Eve in the garden of Eden, thousands of years before the Incarnation. The angel of the Lord, appearing to Abraham, was also Christ. And who do you suppose encountered Moses at the burning bush? Moreover, the infancy narratives of the New Testament speak of angelic visitations - Matthew's angel appears in Joseph's dream, but Luke's angel appears directly to both Zechariah and the virgin Mary. The main biblical apparition of a departed saint appearing to the living involves Samuel and Saul. But, the account of Jesus's death in Matthew XXVII, 52-53, adds another - the tombs of righteous persons were divinely opened and these saints appeared to the living throughout the city of Jerusalem. Wesley noticed all instances of apparitions in the biblical tradition and treated them in his *EXPLANATORY NOTES*.

He also took notice of the history of extra-biblical apparitions in the history of the Church, but he did not treat them as being of equal authority with the scriptural phenomena.

The impact of St. Augustine's theology of apparitions and ghosts on the Catholic Tradition cannot be taken lightly. John Wesley was schooled in this intellectual climate, and his many affirmations on ghosts and apparitions are consistent only within the Augustinian framework.

One of the moral treatises of St. Augustine was entitled *ON CARE TO BE HAD FOR THE DEAD*. Without treating the entire work, a few and important principles must be observed.

1. Wherever the flesh of the departed may lie or not lie, the spirit (soul) requires rest and must get it.

2. The spirit in departing takes with it the consciousness without which it cannot distinguish how one exists, whether in a good estate or in a bad.

3. It does not look for aiding of its life from the flesh to which it withdrew in death.

4. The spirit will return to the resurrected flesh, either to punishment or glory.[109]

5. Until the resurrection, spirits of the righteous dead learn some things that are occurring in the earthly life -

"What things it is necessary that they should know, and what persons it is necessary should know the same, not only things past or present, but even future, by the Spirit of God revealing them: like as not all men, but the Prophets while they lived here did know, nor even they all things, but only what things to be revealed to them the providence of God judged meet."[110]

6. "Some from the dead are sent to the living, as on the other hand, Paul from the living was rapt into Paradise, divine Scripture doth testify." Also -

"For Samuel the Prophet, appearing to Saul when living, predicted even what should befall the king: although some think it was not Samuel himself, that could have been by magical arts evoked, but that some semblance: though the book Ecclesiasticus, which Jesus, son of Sirach, is reputed to have written, and which on account of some resemblance of style is pronounced to be Solomon's, Samuel even when dead did prophesy. But if this book be spoken against from the canon of the Hebrews, (because it is not contained therein,) what shall we say of Moses, whom certainly we read both in Deuteronomy to have died, and in the Gospel to have, together with Elias who died not, appeared unto the living?"[111]

Another important writing pertaining to apparitions and ghosts is Letter CLVIII, written to Augustine by Evodius, Bishop of Uzala, in A.D. 414. Evodius saw that the nature of the soul is the key to understanding the phenomena of ghosts and apparitions. Consequently he posed a series of questions on the soul, requesting Augustine to "correct me where I am mistaken, and teach me what you know that I am desirous to learn."[112] But more important than his questions are some of his accounts of apparitions. (1) His youthful clerk died at twenty-two years of age, following sixteen days of serious illness, during which time he desired greatly to "depart and be with Christ." At the saintly youth's bedside, Evodius witnessed his departure from this life, and the bishop was filled with ecstatic joy - "I think that after leaving his own body he has entered into my spirit, and is there imparting to me a certain fulness of light from his presence, for I am conscious of a joy beyond all measure through his deliverance and safety - indeed it is ineffable."[113] Two days later, the departed youth appeared in a dream of a devout and respected widow. (2) Visitations in sleep, such as in the case of St. Joseph (Matthew I:18ff), are common. Evodius explained -

"In the same manner, therefore, our own friends also who have departed this life before us sometimes come and appear to us in dreams, and speak to us. For I myself remember that Profuturus, and Privatus, and Servilius, holy men who within my recollection were removed by death from our monastery, spoke to me, and that the events of which they spoke came to pass according to their words. Or if it be some other higher spirit that assumes their form and visits our minds, I leave this to the all-seeing eye of Him before whom every thing from the highest to the lowest is uncovered."[114]

(3) Souls, separated from their bodies at death, have some sort of body, capable of sensual experiences -

> "Else how can we explain the fact that very many dead persons have been ob-
> served by day, or by persons awake and walking abroad during the night, to
> pass into houses just as they were wont to do in their lifetime? This I have
> heard not once, but often; and I have also heard it said that in places in which
> dead bodies are interred, and especially in churches, there are commotions
> and prayers which are heard for the most part at a certain time of the night.
> This I remember hearing from more than one: for a certain holy presbyter
> was eye-witness of such an apparition, having observed a multitude of such
> phantoms issuing from the baptistery in bodies full of light, after which he
> heard their prayers in the midst of the church itself. All such things are either
> true, and therefore helpful ...or are mere fables."[115]

Augustine's answer to the letter of Evodius is quite pedantic in spirit and nature (Letter CLIX). He rejected the idea of a departed soul taking another body - "I by no means believe that the soul in departing from the body is accompanied by another body of any kind."[116] Concerning the accounts of apparitions, Augustine claimed difficulty in explaining the "ordinary" facts of human existence - "The more do I shrink from venturing to explain what is extraordinary."[117] So, in place of rational analysis, Augustine recounted an apparition story of his own - not an ordinary event, but extraordinary -

> "You know our brother Gennadius, a physician, known almost to everyone. . .
> who now lives in Carthage. . . and was.. a medical practitioner in Rome. . . a
> man of religious character and of very great benevolence, actively compas-
> sionate and promptly liberal in his care of the poor. . . Nevertheless
> . . . when still a young man. . .(he) had doubts as to whether there was any life
> after death. . . God would in no wise forsake a man so merciful in his disposi-
> tion and conduct. . . (so) there appeared unto him in sleep a youth of re-
> markable appearance and commanding presence, who said to him: 'Follow
> me.' Following him, he came to a city where he began to hear on the right
> hand sounds of a melody so exquisitely sweet as to surpass anything he had
> ever heard. When he inquired what it was, his guide said: 'It is the hymn of
> the blessed and the holy.' What he reported himself to have seen on the left
> hand escapes my remembrance. He awoke; the dream vanished, and he

thought of it as only a dream. On a second night, however, the same youth appeared to Gennadius, and asked whether he recognized him, to which he replied that he knew him well, without the slightest uncertainty. Thereupon he asked Gennadius where he had become acquainted with him. There also his memory failed him not as to the proper reply: he narrated the whole vision, and the hymns of the saints which, under his guidance, he had been taken to hear, with all the readiness natural to recollection of some very recent experience. On this the youth inquired whether it was in sleep or when awake that he had seen what he had just narrated. Gennadius answered: 'In sleep.' The youth then said: 'You remember it well; it is true that you saw these things in sleep, but I would have you know that even now you are seeing in sleep.' Hearing this, Gennadius was persuaded of its truth, and in his reply declared that he believed it. Then his teacher went on to say: 'Where is your body now?' He answered: 'In my bed.' 'Do you know,' said the youth, 'that the eyes in this body of yours are now bound and closed, and at rest, and that with these eyes you are seeing nothing?' He answered: 'I know it.' 'What, then,' said the youth, 'are the eyes with which you see me?' He, unable to discover what to answer to this, was silent. While he hesitated, the youth unfolded to him what he was endeavoring to teach him by these questions, and forthwith said: 'As while you are asleep and lying on your bed these eyes of your body are now unemployed and doing nothing, and yet you have eyes with which you behold me, and enjoy this vision, so, after your death, while your bodily eyes shall be wholly inactive, there shall be in you a life by which you shall still live, and a faculty of perception by which you shall still perceive. Beware, therefore, after this of harbouring doubts as to whether the life of man shall continue after death.' This believer says that by this means all doubts as to this matter were removed from him. By whom was he taught this but by the merciful, providential care of God?"[118]

Augustine's doctrine of the soul, therefore, with its emphasis on being created immortal with a set of spiritual sense perceptions that function effectively after the soul has departed the body in death, gives a strong foundation for explaining the historic tradition of apparitions and ghosts. John Wesley clearly stood in this theological context of the soul.[119] His doctrine of the senses of the soul,[120] and their employment after death, was built upon this Augustinian foundation, making belief in ghosts and apparitions quite theological.

Wesley read other works on this subject. In December 1764, while returning to London from the southern part of England, he read *THE CERTAINTY OF THE WORLD OF SPIRITS* by Richard Baxter, published in 1691, and now enshrined in the archives of the Wesley Historical Society, Volume IV, 137. It will be remembered that Wesley regarded Baxter as one of the greatest Christian authorities on Scripture, Christian doctrine, and piety. Moreover, Wesley included several of Baxter's writings in his *CHRISTIAN LIBRARY* for his Methodists to read and study. Of Baxter's work on apparitions, Wesley said - "It contains several well-attested accounts; but there are some which I cannot subscribe to. How hard is it to keep the middle way; not to believe too little or too much."[121] Six days later, Wesley conducted the funeral service of Mrs. Prior, the housekeeper of Mr. P____ (Parker?). Mr. P. related the immediate aftermath of her death - At one a.m., he rang for the butler. Upon his coming into the bedroom, Mr. P. said to him, "Mrs. Prior is dead. She just now came into my room, and walked round my bed." About two, the nurse came to Mr. P.'s room and told of Mrs. Prior's death. Mr. P. inquired as to the time of her death. The nurse replied, "Just at one o'clock."[122] Wesley apparently had little difficulty in believing this account. It was well within the "middle way" of faith and reason.

However, Wesley was swimming up stream, against a strong flow of Enlightenment thought. Ghosts, apparitions, and witchcraft, according to the progressive thinkers of the age, were non-existent, the figments of ignorant minds, created by unscrupulous religionists to control through fear the lives of the masses. Having argued from the biblical tradition and the Patristic tradition of St. Augustine, which the Enlightenment generally rejected, Wesley turned to reason in support of his belief in ghosts and apparitions. His formal arguments took the voice of "solemn protest."

"The English in general, and indeed most of the men of learning in Europe, have given up all accounts of witches and apparitions, as mere old wives' fables. I am sorry for it; and I am willing to take this opportunity of entering my solemn protest against this violent compliment which so many that believe the Bible pay to those who do not believe it. I owe them no such service. I take knowledge these are at the bottom of the outcry which has been raised, and with such insolence spread throughout the nation, in direct opposition, not only to the Bible, but to the suffrage of the wisest and best men in all ages and nations. They well know (whether Christians know it or not), that the giving up witchcraft is, in effect, giving up the Bible; and they know, on the other hand, that if but one account of the intercourse of men with separate spirits be admitted, their whole castle in the air (Deism, Atheism, Materialism) falls to the ground. I know no reason, therefore, why we should suffer even this weapon to be wrested out of our hands. Indeed there are numerous arguments besides, which abundantly confute their vain imaginations. But we need not be hooted out of one; neither reason nor religion require this."[123]

Following this introduction, Wesley launched into rational answers to objections raised by unbelieving critics. (1) The chief objection to the reality of ghosts and apparitions is put in the form of a question - Did you ever see an apparition yourself? Wesley's answer - "No; nor did I ever see a murder; yet I believe there is such a thing; yea, and that in one place or another murder is committed every day." His conclusion was precise - "Therefore I cannot, as a reasonable man, deny the fact, although I never saw it, and perhaps never may. The testimony of unexceptionable witnesses fully convinces me both of the one and the other."[124] Reason dictated for Wesley that all claims to apparitional experience be tested by the credibility of its witnesses. As he recounted many stories of ghosts and strange apparitions, Wesley invariably treated the reliability of the witnesses. In most cases, he tried to interrogate the witnesses to extraordinary phenomena of this sort, if at all possible. (2) Another objection of the critics, treated by Wesley, claimed that all alleged preternatural operations are the inventions of "artful men" and nothing more. Apparitions are actually tricks, the result of clever manipulation of simple minds with man-made devices, mirrors, etc. Wesley quickly recognized that there are many tricksters in the area, but clever men are not the cause of every instance of this type. To illustrate this principle, he cited the famous episode of the "drumming" in Mr. Mompesson's house at Tedworth. Famous indeed! Wesley included the account in the *ARMINIAN*

MAGAZINE (1785). Earlier, in 1688, the account was given under the title, *NARRATIVE OF THE DEMON OF TEDWORTH*, or *OF THE DISTURBANCES AT MR. MOMPESSON'S HOUSE*. Addison used this story for his comedy, *THE DRUMMER*. Hogarth's cartoon, *CREDULITY, SUPERSTITION, AND FANATICISM: A MEDLEY*, depicted a thermometer rising from a Methodist's brain, and at the top of the print was the word "Tedworth" and a drummer drawn small over "Cock Lane Ghost."[125] What a glorious time the critics of early Methodism had because of Wesley's identification of the drummer affair as a true apparition. Now, in arguing with the critics, Wesley included the drummer in his rational attempt to advance the cause of believers. The opposition claimed that, after a long search through his house, Mr. Mompesson found a "contrivance" used by some trickster to produce the mysterious and nocturnal drumming. To this claim Wesley retorted -

> "Not so; my eldest brother, then at Christ Church, Oxon, inquired of Mr. Mompesson, his fellow collegian, whether his father had acknowledged this or not. He answered, 'The resort of gentlemen to my father's house was so great he could not bear the expense. He therefore took no pains to confute the report that he had found out the cheat; although he, and I, and all the family, knew the account which was published to be punctually true.'"[126]

(3) Declaring the two preceding arguments to be "premises" - as if he were constructing an Aristotelian syllogism - Wesley moved to draw a deductive conclusion from the form and content of these premises. The conclusion, however, took the shape of "a remarkable narrative" that featured a lengthy account of apparitions to Elizabeth Hopson. He began with a word to his readers - "The reader may believe it if he pleases; or may disbelieve it, without any offence to me. Meantime, let him not be offended if I believe it, till I see better reason to the contrary."[127] The story runs: Elizabeth was born in Sunderland in 1744, her father dying when she was three or four years of age. Thomas Rea, her uncle, raised her as his own daughter. As a child, she was serious, filled with the fear of God and a deep sense of sin. When sixteen she found peace with God. In May of 1768, Wesley visited her and interrogated her

for four days concerning her frequent encounters with ghosts. Her account to Wesley included the following experiences - (a) As a child, when any of her neighbors died, "I used to see them, either just when they died, or a little before." She claimed not to be frightened by them because "it was so common." Often, though, she did not know that they were dead. These appearances occurred in broad daylight, and many at night. The ghosts coming in the dark brought light with them. "All little children, and many grown persons, had a bright, glorious light around them" But, other persons - obviously sinners in the former life - "had a gloomy, dismal light, and a dusky cloud over them." Her uncle, upon learning of her special gift, assured her that no harm could come to her as long as her trust in God remained strong during these apparitions. He also informed her that "evil spirits seldom appear but between eleven at night and two in the morning; but after they have appeared to a person for a year, they frequently come in the day-time. Whatever spirits, good or bad, come in the day, they came at sunrise, at noon, or at sunset."[128] (b) Elizabeth, nearly thirteen years of age, next encountered the spirit of William, an evil lodger in her uncle's house, while his body was fast asleep in his room. Several days later he died "in raging despair."[129] (c) At fifteen years of age, Elizabeth was sent early one morning to get the cows from the field, but this necessitated crossing two other fields, both of which were reputed to be haunted. She had many times passed through this haunt, hearing the sounds of quarrelling and seeing people materialize and then vanish before her eyes. On this morning, however, as she neared the gate of the field where the cows were, suddenly a young man dressed in purple appeared on the other side, and addressed her - "It is too early; go back from whence you came. The Lord be with you, and bless you." Then he vanished.[130] (d) When Elizabeth was sixteen, her uncle took ill and declined nearly to the point of death after several months. One day, returning home from fulfilling an errand, Elizabeth saw him coming swiftly across the field toward her. She ran to meet him but suddenly he vanished. When she entered the house, he was in bed, calling for her. She sank into his arms, and with many tears he admonished her to "continue in the ways of God." He then "sunk down and died."[131]   (e) Elizabeth fell into a long state of mourning, weeping both day and night, until her body was weakened, being itself near death. One morning, about one o'clock, as she laid in her bed crying, she heard a noise. Rising up, she saw the ghost of her uncle at her bedside. He looked displeased, shook his head at her, and after a few minutes he vanished. A week later she took to the bed completely, intending to die there. Six nights later, at about eleven o'clock, her uncle's ghost came into her room once more, looked at her, seeming to be pleased, and sat on the bedside. "He came every night after, at the same time, and stayed till cockcrowing." Elizabeth kept her eyes fixed upon him the entire time of his visitations. Whenever she needed a drink of water, or something else, although she did not ask, "he fetched it, and set it on the chair by the bedside." Elizabeth could not speak to him, her tongue seeming to be bound. When he departed at dawn, he waved good-bye, and Elizabeth heard delightful music, "as if many persons were singing together."[132]   (f) Six weeks later, she was stronger in body and mind. One night, as she laid in bed, praying and musing concerning her uncle, he appeared again, dressed now in a white robe, and he looked very pleased. Suddenly there stood beside him another robed person, tall and quite beautiful in countenance. The room was then filled with singing until the crowing of the cock. The uncle now smiled, waving his hand in departing gesture, and "they went away with inexpressibly sweet music" to be seen no more by Elizabeth Hopson.[133]   (g) The next year Elizabeth was courted by a young seaman, and they planned to marry after he returned from his next voyage. He boarded his ship one evening, preparing to sail early in the morning. But he appeared to Elizabeth that night, she not knowing that it was an apparition. Reaching out to touch him, he fled from her, passing through a wall. The next morning at ten o'clock, he died. (h) A few days later, John Simpson, a neighbor, went to sea on Tuesday, but on the following Friday night, between eleven and twelve, Elizabeth was awakened by the sound of someone walking in water in her room. "He then came to the bedside, in his sea-jacket, all wet, and stretched his hand over me. Three drops of water fell on my breast and felt cold as ice. I strove to wake his wife, who lay with me; but I could not." In a short while he went away. But he returned every night for a week, between eleven and two. Before he came, and after he departed, sweet music flooded the room. From this time onward, the ghost of John Simpson came both day and night - "every night about twelve, with the music at his coming and going, and every day at sunrise, noon, and sunset." He appeared to Elizabeth at church, at the Methodist preaching house, at the Methodist class meeting, "and was always just before me, changing his posture as I changed mine." When she sat, he sat. When she knelt in prayer, so did he. For ten weeks he continued his appearing to her, and she declined in health and mental well-being. He then came four or five consecutive nights to her room, but without any music. On the next night, he came into the bedroom and violently shook the curtains around the bed, "looking wistfully" at Elizabeth. This was repeated over the next few nights, until she fled into the garret of the house. He followed her and began to close the distance between them. She cried out to him, "In the name of the Father, Son, and Holy Ghost, what is your business with me?" He answered, "Betsy, God forgive you for keeping me so long from my rest. Have you forgot what you promised before I went to sea - to look to my children, if I was drowned? You must stand to your word, or I cannot rest." She retorted, "I wish I was dead." He said, "Say not so; you have more to go through before then; and yet, if you knew as much as I do, you would not care how soon you died. You may bring the children on in their learning, while they live; they have but a short time to live." She promised, "I will take all the care I can." He warned her not to accept the invitation of her brother to come to live with him in Jamaica - an invitation that she had not yet received through the post. He warned her of entering into marriage with a man she had been considering, and he charged her to be true to God in all things. She then asked, "How do you spend your time?" "In songs of praise. . . I have lost much happiness by coming to you. . . The Lord would not suffer me to fright you. . . I shall only come to you twice more before the death of my two children. God bless you."

The room was filled with singing of a thousand voices or more as Elizabeth followed him downstairs. She asked for him to return and then posed several more questions. He responded, "I wish you had not called me back, for now I must take something from you. . . I think you can best part with the hearing of your left ear." His hand touched the ear and it became stone deaf. The cock crowed and he departed, the music ceased. The older child died when he was three and a half years old, the younger before he was five. John appeared to Elizabeth Hopson before the death of each child, but he did not speak to her. Afterwards, she saw him no more.[134]

Here is an appropriate place to terminate treatment of the Hopson account, although we have examined only half of the story. It would be worth the reader's time and effort to examine the entire account. Nehemiah Curnock's footnote to the account[135] should not be overlooked, in which a Quaker's attack upon both the story and Wesley is offered. The Hopson story, however, is just one of many accounts included by Wesley in his writings. A brief comment about other accounts is certainly in order before concluding this section of our study.

At Loddon, ten miles from Norwich, Wesley interrogated four persons of a house concerning a strange occurrence. These four, two men and two women, were of good sense and impeccable character, all possessing a great fear of God. Wesley held them to be completely credible. The essentials of their account are - The two men were told by another that a treasure was hidden in a spot near Norwich, consisting of money and plate in a chest, six to eight feet below the surface of the earth. They did not believe this report, but at night each man had the same apparition - an elderly man and woman stood by their beds, and commanded them to go and dig up the treasure between eight and midnight. Being afraid, they took a third man and began digging. At six feet they arrived at the chest, but it began to sink into the earth. Immediately a large globe of bright fire hovered over the place, then moved upward until it passed from sight. The elderly man and woman appeared and chided the two men - "You spoiled all by bringing that man with you." From that time onward the two men and the two women of the house heard, several times a week, "delightful music for a quarter of an hour at a time. They often hear it before those persons appear; often when they do not appear."[136]

Another series of apparitions was told to Wesley by a trusted friend of years, Mrs. Sarah Maitland, whom he trusted greatly. Three different episodes comprise this account: (1) Her husband's servant died. Several days later, as she was walking into town, she met him in his everyday clothes running towards her. In about a minute, he disappeared. (2) A surgeon and apothecary, Mr. Heth, died in March. In April, Mrs. Maitland and several other women were walking on High Street at daybreak, and they all three saw him, dressed as usual with a small hat perched on his bushy wig. As they approached him, he silently passed by them and turned into the market-house. And (3) Ten or twelve days after Mr. Smith's death, Mrs. Maitland passed his house and observed him standing at his chamber window, looking out at her. He possessed a most horrid countenance, and soon withdrew from the window.[137] Wesley offered no commentary.

A young woman, aged twenty-six, living near Colne where Wesley often preached, and a person of "unblameable behaviour," related a rather standard apparition tale, which Wesley believed and recorded in his JOURNAL. The essence of it - At the age of four, she began seeing ghosts walking in her bedroom. They would come close to her bed, look down upon her but not speak. Some of them looked very sad and others looked very happy. Two of them were husband and wife from the parish church, a couple known for their fighting. At last one ghost, a sixteen year old boy who had died the week before, spoke to the young girl. She asked, "You are dead! How did you get out of the other place?" He explained how easy it is, but she offended him by an inappropriate remark. He grew angry and frightened her, causing her to pray. He quickly disappeared. The others continued to appear in her room at night, having an aura of light always about them. These apparitions continued until she was sixteen, at which time she started to be troubled by them. The towns people began to talk against her, calling her a witch. "This made me cry earnestly to God to take it away from me. In a week or two it was all at an end; and I have seen nothing since."[138]

One of Wesley's ghost stories contained the theme that the departed spirit of Richard Sims was sent to Daniel Car to commission him to visit the ghost's nephew and niece and warn them to turn to God. The nephew, Thomas Roberts, "will die on the 26th of next month, and she will die on the 30th." When Car objected to going, the ghost ordered him to take pen and paper and write what he dictated - a description of hell and its torments, "in such words as I had never heard in my life, enough to make one's blood run cold." The ghost of Richard Sims then ordered Car to pick up his Bible and open it, reading the first words his eyes fell upon. He did this and read from John V, 28 and 29 - "The hour is coming in which all that are in the graves shall hear His voice, and shall come forth; they that have done good unto the resurrection of life, and they that have done evil unto the resurrection of damnation." Car concluded his narration with - "At that word he gave such a groan and shriek as I never heard and. . ."[139] The account was left unfinished by Wesley, probably for dramatic purposes, to give the readers' minds something to ponder and imagine.

An unusual apparition, reported by Wesley, involved a woman, with an open Bible before her, entertaining a ghost whom she thought to be her uncle, but was not. She suspected as much because of his speech, and she observed that one of his feet was like the foot of an ox. He terrorized her until she agreed to kill her father. She tried to take her own life on several occasions, attempting to avoid murdering her father, but always without success. Her brother placed her in a straight-jacket, "such as they use in Bedlam." But, she escaped her confinement and was found up a chimney. More elaborate straps were added to the jacket. However, as she laughed, they snapped and she was free once more. A physician was called who announced that her condition was "partly natural, partly diabolical," and that prayer alone was the remedy. Finally, the story came to a happy, dramatic ending, with Jesus claiming her soul and body, snatching them from the devil, enabling her to sing lustily -

"O Sun of Righteousness, arise,
    With healing in Thy wing;
To my diseased, my fainting soul,
    Life and salvation bring."[140]

A surprising apparition to a devout young woman, as she de-scribed it, has strong feminist overtones - a female angel frequently came to her. The report interested Wesley greatly, and he ques-tioned her for an entire day in June 1788. A portion of Margaret Barlow's account reads -

"For about a year, I have seen this angel, whose face is exceeding beautiful; her raiment (so she speaks) white as snow, and glistening like silver; her voice unspeakably soft and musical. She tells me many things before they come to pass. She foretold I should be ill at such a time, in such a manner, and well at such an hour; and it was so exactly. She said, such a person shall die at such a time; and he did so. Above two months ago she told me your brother was dead (I did not know you had a brother); and that he was in heaven. And sometime since she told me you will die in less than a year. But what she has most earnestly and frequently told me is that God will in a short time be avenged of obstinate sinners, and will destroy them with fire from heaven."[141]

Wesley's reaction to Margaret's account was mixed. On one hand, he was uncertain that it could ever be verified - "Whether this be so or no, I cannot tell." On the other hand, he was greatly impressed with "the wonderful power in her words. . .They did good to my heart." What Wesley regarded as incredible was that the an-gel in question was a female. This was, he felt, a mistake - ". . . from the face, the voice, and the apparel, she might easily mistake *him* for a female." Nevertheless, he continued, "this mistake is of little consequence."[142] Perhaps in his day it was of little consequence, but not in the contemporary age! However, she was surely wrong about the year of Mr. Wesley's death - He lived the entire year of 1788 plus three more.

No treatment of Wesley's ghosts and apparitions should exclude mention of Old Jeffrey. Nearly everyone who reads the history of early Methodism knows that Old Jeffrey was the parsonage ghost at Epworth. Although he was not at home when the ghost caused remarkable disturbances, being at Christ Church at the time, his family kept him informed. In March of 1784, Wesley reduced the apparitions to narrative form for his *ARMINIAN MAGAZINE*.[143] Umphrey Lee identified Old Jeffrey as a Jacobite ghost.[144] Being a Jacobite was synonymous with being a Non-juror, and giving complete loyalty to James II who abdicated the throne of England in 1688.[145] One of the best descriptions of the Wesley family's response to Old Jeffrey is that of Bishop McConnell -

"Did they seek to move out of the house and abandon it as haunted? They did not. No ghost, real or imagined, could scare, after the first few disturbances, the Wesley children. They took the noises as they came - tried at first to get into communication with the spirit, and finally treated 'him' with an easy familiarity bordering on amusement. Samuel tried to devise a scheme of return knocks to the tappings, with apparently as cool a rationalism as if he belonged to a psychical research society of our modern type. The girls used to believe that they felt Old Jeffrey holding a door shut when they tried to open it. Instead of running and screaming in terror, they pulled harder on the door, and when they got it free indulged in ejaculations of contemptuous defiance at the rude visitor from the realms of shade."[146]

Notes

1 EXPLANATORY NOTES OLD TESTAMENT, *op.cit.*, I, 26.
2 Epistle to the Romans, V, 12ff.
3 EXPLANATORY NOTES NEW TESTAMENT, *op.cit.*, 537.
4 *Ibid*, 538.
5 *Ibid*, 538, 634-635 - "For as through Adam all die, even so through Christ shall all be made alive."
6 *Ibid*, 635.
7 *Loc.cit.*
8 EXPLANATORY NOTES OLD TESTAMENT, *op.cit.*, III, 2492.
9 *Loc.cit.*
10 WORKS, *op.cit.*, VI, 496.
11 *Loc.cit.*
12 *CERTAIN SERMONS OR HOMILIES TO BE READ IN CHURCHES IN THE TIME OF QUEEN ELIZABETH OF FAMOUS MEMORY* (London: SPCK, 1843), 96-97. Cf. WORKS, *op.cit.*, VII, 204 - Wesley regarded the HOMILIES to be "next to Scripture."
13 *Ibid*, 96.
14 WORKS, *op.cit.*, VII, 327.
15 *Loc.cit.*
16 *Loc.cit.*
17 *Loc.cit.*
18 EXPLANATORY NOTES NEW TESTAMENT, *op.cit.*, 266.
19 *Ibid*, 266-267.
20 *Ibid*, 267.
21 *Loc.cit.*
22 *Loc.cit.*
23 WORKS, *op.cit.*, VII, 327; VI, 496.
24 CERTAIN SERMONS, *op.cit.*, 99.
25 WORKS, *op.cit.*, X, 98, 145; VII, 247. Wesley rejected the Roman Catholic division of hell into *Limbus Patrum*, *Limbus Infantum*, Purgatory, and the Damned.
26 *Ibid*, X, 99.
27 *Loc.cit.*
28 *Ibid*, VII, 246.
29 *Loc.cit.* Cf. Telford, *op.cit.*, VI, 213-214. In this letter, dated April 17, 1776, Wesley made the same distinction between Paradise and heaven, referring to Paradise as the "Intermediate State."
30 *Ibid*, VII, 247.
31 *Ibid*, VII, 252.

32 EXPLANATORY NOTES NEW TESTAMENT, *op.cit.*, 31-32.
33 WORKS, *op.cit.*, VII, 68.
34 Jeremy Taylor, *HOLY LIVING AND HOLY DYING* (New York: Appleton Company, 1847), 42.
35 *Ibid*, 43.
36 *Ibid*, 45.
37 *Ibid*, 47.
38 *Ibid*, 54.
39 EXPLANATORY NOTES OLD TESTAMENT, *op.cit.*, I, 191.
40 WORKS, *op.cit.*, XII, 114.
41 *Ibid*, VI, 496.
42 *Loc.cit.*
43 *Ibid*, VI, 384.
44 *Ibid*, VI, 496-497.
45 *Ibid*, VII, 328. Cf. William Beveridge, *THEOLOGICAL WORKS* (Oxford: John Bishop of St. Asaph's, 1842), I, 375. Beveridge described Lazarus and the other saints in Abraham's bosom "swimming in the rivers of pleasure" while Dives could not have one drop of water on his tongue.
46 Tyerman, *op.cit.*, III, 158. Cf. WORKS, *op.cit.*, VII, 332. Cf. Thomas Ken. *op.cit.*, 140-141.
47 WORKS, *op.cit.*, XII, 344-345.
48 *Ibid*, VI, 383.
49 Telford, *op.cit.*, VII, 168.
50 *Ibid*, VI, 213.
51 WORKS, *op.cit.*, VI, 384.
52 *Ibid,,* VII, 328.
53 *Loc.cit.*
54 JOURNAL, *op.cit.*, III, 197-198.
55 *Ibid*, V, 191.
56 *Ibid*, V, 299-300.
57 Prince, *op.cit.*, 129.
58 JOURNAL, *op.cit.*, V, 400, 306.
59 *Ibid*, VI, 392 note.
60 *Ibid*, VI, 392.
61 *Ibid*, VI, 255.
62 *Ibid*, VI, 90, 255, 392.
63 CERTAIN SERMONS, *op.cit.*, 95-96.
64 JOURNAL, *op.cit.*, I, 123.
65 *Ibid*, I, 138.
66 *Ibid*, I, 139. Wesley deliberately nursed his fear of death in order to become a true Christian without fear - a self-torture approach.
67 *Ibid*, I, 140.
68 *Ibid*, I, 110, 141.
69 *Ibid*, I, 141-142.
70 *Ibid*, I, 143.
71 *Ibid*, I, 417.

72   *Ibid*, I, 418.
73   *Ibid*, I, 423.
74   *Ibid*, I, 435.
75   *Ibid*, I 471.
76   *Ibid*, III, 91-92.
77   *Ibid*, IV, 90-91.
78   *Ibid*, VIII, 143-144.
79   *Ibid*, VIII, 144. Wesley was buried in the yard behind the New Chapel on City Road in London. His wife was buried in the churchyard of Camberwell in London.
80   *Ibid*, I, 139.
81   *Ibid*, V, 45, 314, 316, 319; VI, 19, 29, 40 , 304, 341.
82   WORKS, *op.cit.*, VI, 381.
83   *Loc.cit.*
84   *Ibid*, VI, 382.
85   *Loc.cit.*
86   *Loc.cit.*
87   *Ibid*, VI, 383.
88   *Ibid*, VI, 384.
89   *Ibid*, VI, 385-386.
90   *Ibid*, VI, 386.
91   *Loc.cit.*
92   *Ibid*, VI, 387. Cf. NICENE FATHERS, *op.cit.*, II, 452-454. Augustine used the same essential argument.
93   *Loc.cit.*
94   *Ibid*, VI, 388.
95   *Ibid*, VI, 389.
96   *Loc.cit.*
97   *Ibid*, VI, 390.
98   *Ibid*, VI, 391.
99   *Ibid*, IX, 508.
100 *Ibid*, VI, 497; VII, 234.
101 *Ibid*, VII, 234.
102 *Ibid*, VI, 497. Cf. METHODIST HYMNAL, *op.cit.*, 811.
103 McConnell, *op.cit.*, 289-290.
104 EXPLANATORY NOTES OLD TESTAMENT, *op.cit.*, II, 985.
105 *Loc.cit.*
106 *Loc.cit.*
107 *INTERPRETER'S DICTIONARY OF THE BIBLE* (Nashville: Abingdon Press), IV, 302.
108 EXPLANATORY NOTES OLD TESTAMENT, *op.cit.*, II, 1581.
109 NICENE FATHERS, *op.cit.*, III, 543.
110 *Ibid*, III, 548.
111 *Loc.cit.*
112 *Ibid*, I, 513.
113 *Ibid*, I, 509.

114 *Ibid*, I, 512.
115 *Ibid*, I, 511.
116 *Ibid*, I, 513; V, 332. Augustine held that the soul has spiritual senses as the body has physical senses.
117 *Loc.cit.*
118 *Ibid*, I, 514; V, 332.
119 WORKS, *op.cit.*, VI, 137, 140, 191, 343, 395; VII, 70, 226, 227, 320, 329, 332, 371.
120 *Ibid*, VI, 70, 384.
121 JOURNAL, *op.cit.*, V, 103.
122 *Loc.cit.*
123 *Ibid*, V, 265. Cf. Lee, *op.cit.*, 37. The death penalty for witchcraft in England was abolished in 1736, indicating the flow away from church tradition.
124 *Ibid*, V, 266.
125 *Loc.cit.*
126 *Loc.cit.*
127 *Loc.cit.*
128 *Ibid*, V, 267.
129 *Loc.cit.*
130 *Ibid*, V, 268.
131 *Loc.cit.*
132 *Loc.cit.*
133 *Ibid*, V, 269.
134 *Ibid*, v, 269. The reference to a thousand voices, singing praise to God, may possibly be the origin of Charles Wesley's hymn, O For a Thousand Tongues to Sing My Great Redeemer's Praise.
135 *Ibid*, V, 295.
136 *Ibid*, V, 487.
137 *Ibid*, VI, 212-213.
138 *Ibid*, V, 178.
139 *Ibid*, IV, 148-149.
140 *Ibid*, V, 32-35.
141 *Ibid*, VII, 398-399.
142 *Ibid*, VII, 399.
143 *Ibid*, VI, 487-488. Cf. Lee, *op.cit.*, 49. Lee claims Wesley was at Charterhouse at the time. Curnock is correct in placing him at Oxford.
144 Lee, *op.cit.*, 49.
145 David I. Naglee, *FROM FONT TO FAITH: JOHN WESLEY ON INFANT BAPTISM AND THE NURTURE OF CHILDREN* (New York, Berne: Peter Lang Company, 1987), 8.
146 McConnell, *op.cit.*, 15.

Part Three

ETERNITY: A PARTE POST
(After Time)

Chapter Nine

The End and the Beginning: The Signs of Christ's Coming

"When the Son of Man shall come in his glory and assign every man his own reward, that reward will undoubtedly be proportioned, first to our inward holiness or likeness to God, secondly to our works, and thirdly to our sufferings; therefore for whatever you suffer in time, you will be an unspeakable gainer in eternity. Many of your sufferings, perhaps the greatest part, are now past; but the joy is to come! Look up, my dear friend, look up and see the Crown before you! A little longer, and you shall drink of the river of pleasure that flows at God's right hand for evermore!"[1]

These triumphant words, written to Ann Bolton by John Wesley on December 15, 1786, testify to his extraordinary anticipation of the nearness of the end of time and the beginning of eternity *a parte post*. While many persons fear the end of time, Wesley welcomed it. The "end" was not cataclysmic disaster spawned by divine wrath, but, rather, for Wesley, it was an orderly progression of divine grace and justice that arrives at a teleological terminus, at a planned goal of perfection. It is the "end" to which all creation, from the beginning of time, has been moving according to God's eternal plan. The world order was made good at creation, then it was made subject to vanity by man's sin, but God's plan to bring the good to eternal perfection could not be thwarted by sin. Where sin abounded, Wesley argued, divine grace did much more abound. In spite of sin and evil, God kept to his plan of a *telos* when time will cease and a truly perfect Paradise will enter eternity. Such an "end" is to be welcomed. It is an optimistic hope, stressing the good - though afflicted with sin and evil - becoming altogether perfect. No one should accuse Wesley of being pessimistic in his eschatological views. He did not believe for a instant that the present world-order was so corrupted that God would have to destroy it entirely and substitute a new creation in its place. Actually, he believed, as we shall see, that the created stuff of the world-order gets purified by fire during the last days of time, and that this created matter becomes the New Creation. Matter is not destroyed because it has been made indestructible, being divinely destined for eternal existence. Moreover, in his eschatological optimism, the Christian pilgrim joyously awaits this glorious end, serving Christ with untiring zeal, while his spiritual eyes look up, clinging to heaven, from whence the living Christ will come with angels and the souls of the saintly dead, initiating the last things which constitute the "end."

SIGNS OF THE END FROM GOSPEL AND EPISTLE

Young Timothy had been warned - "In the last days grievous times will come" (II Timothy III, 1). Wesley understood "the last days" to be the "gospel dispensation, commencing at the time of our Lord's death."[2] While it is marked by grievous times, it is, nevertheless, a dispensation of great grace, continuing until the end of time.[3] Already, Wesley held, the "last days" have endured for nearly eighteen centuries, but they may end in the year 1836.[4] During these last days, there have been many signs pointing to the end - signs as events of history, indicating the progression of God's plan and Satanic opposition to it, with God holding the upper hand. As the latter days near their end, Satan fights furiously to save his tottering kingdom, but to no avail!

Wesley spoke of many signs of the end, some extracted from the scriptures and others drawn from the history of the Christian Church. A survey of the biblical sources used in his doctrine of the end things is essential to an understanding of Wesley's eschatology. He attempted to be true to the scriptural development of the theme in the New Testament. Believing that the Gospel of Matthew was the first written of the four, he gave the apocalyptic passage of chapter XXIV a thorough examination in his *EXPLANATORY NOTES UPON THE NEW TESTAMENT*. The original teachings of Jesus concerning the signs of the end, he believed, are contained in this passage, while Mark and Luke altered and reduced them.

In Matthew XXIV, 1ff, Wesley noted, Jesus left the temple in Jerusalem and crossed over the Kidron Valley to the Mount of Olives. From the Mount, looking westward, he and his disciples saw the Herodian temple in its entirety. The disciples then asked when the temple would be destroyed, what the signs of Christ's coming were, and what the signs of the end of the world (age) would be. Wesley saw the remainder of this chapter as Christ's answers to these questions, the last two being one and the same.

A. *The Destruction of the Temple* - Wesley argued that verses 4 through 22 represent the answer of Jesus on this subject. The earlier prediction of Jesus (verse 2), that not one stone of the temple would be left standing on another, was fulfilled by Titus, the Roman general who captured Jerusalem in A.D. 70. Wesley, following the account of Flavius Josephus, reported - "This was punctually fulfilled; for after the temple was burned, Titus. . . ordered the very foundations of it to be dug up; after which the ground on which it stood was ploughed up by Turnus Rufus."[5] The time of the temple's destruction was predicted by Christ forty years before. But, Wesley maintained, there were "signs" that preceded the fulfillment - signs related to the destruction of the temple and city - (1) Many false christs and prophets shall arise before then, and again later before the coming of Christ. "And, indeed, never did so many impostors appear in the world as a few years before the destruction of Jerusalem."[6] (2) "Wars" near Jerusalem, and "rumours of wars" at a distance from Jerusalem. "But the end (telos) is not yet."[7] These signs are the beginning of sorrows for the Christians of Jerusalem. They shall be held responsible for the evils befalling the city, and they shall be hated of all nations. (3) Some of these Christians, being offended by such injustice, shall make shipwreck of their faith, but those who truly hope "shall be snatched out of the burning."[8] (4) The "love of many "Christians within the city will "wax cold." (5) Before the destruction of the temple, the gospel will be preached throughout the world, not universally (everywhere) but in particular places. "And this was done by St. Paul and the other apostles before Jerusalem was destroyed."[9] And then shall the end of Jerusalem and the temple come. Wesley added, "Josephus's *HISTORY OF THE JEWISH WARS* is the best commentary on this chapter. It is a wonderful instance of God's providence, that he, an eyewitness, and one who lived and died a Jew. . . should. . . so exactly illustrate this glorious prophecy, in almost every circumstance."[10]   (6) The Roman standards being brought into the holy city by the soldiers constitutes another sign of the nearness to the end of the temple and the city. When brought into the holy place of the temple, the pagan images on the standards certainly represent the "abomination of desolation" foretold by the prophet Daniel.[11] When all these signs appear, without any exception, the inhabitants of Jerusalem should flee to the mountains surrounding the city - "So the Christians did and were preserved."[12] As the city is being attacked by the Romans, let the one on a housetop remain there in hiding. Woe to pregnant women in that day. "Pray that your flight be not in the winter - They did so; and their flight was in the spring."[13] With the fall of the city and the destruction of the temple, "then shall be great tribulation, such as was not from the beginning of the world." Wesley correctly placed this "Great Tribulation" in the historical context of the destruction of the temple and not at the end of time. He left it where Jesus put it! With the complete subduing of the city by the army, the great tribulation shall be shortened and many lives spared for "the elects' sake" - for the sake of the Christians still within the city.[14]

B. *The Signs of Christ's Coming* - Added to this theme, Wesley noted, is the subject of the end of the world or age (verses 23 -51). The signs heralding the *Parousia* of Christ follow the destruction of the temple in Jerusalem - (1) False christs and prophets arise and perform great signs and wonders, deceiving even the elect Christians, but the true Christ will not be among them. When Christ does come, his appearance will be as swift as the lightning from the east that suddenly lights up the entire sky.[15] The false christs and prophets will then become as the carcass eaten by the eagles - the eagles of Rome eating the false saviors of Judaism.[16] (2) Immediately after the tribulation of the temple's destruction, another sign shall appear - "the sun shall be darkened, and the moon shall not give her light, and the stars shall fall from heaven, and the powers of the heavens shall be shaken." Wesley added a special commentary to this description of the second sign -

"Here our Lord begins to speak of His last coming. But He speaks not so much in the language of man as of God, with whom a thousand years are as one day, one moment. Many of the primitive Christians, not observing this, thought He would come immediately, in the common sense of the word: a mistake which St. Paul labours to remove in his Second Epistle to the Thessalonians."[17]

Before leaving this second sign, Wesley treated another New Testament passage which elaborates the theme of one day being as a thousand years with God (II Peter III, 8ff) -

"That one day is with the Lord as a thousand years, and a thousand years as one day - Moses had said, Ps. XC, 4, 'A thousand years in Thy sight are as one day;' which St. Peter applies with regard to the last day, so as to denote both His eternity, whereby He exceeds all measure of time in His essence and in His operation; His knowledge, to which all things past or to come are present every moment; His power, which needs no long delay, in order to bring its work to perfection; and His longsuffering, which excludes all impatience of expectation, and desire of making haste. One day is with the Lord as a thousand years - That is, in one day, in one moment, He can do the work of a thousand years. Therefore He is not slow: He is always equally ready to fulfill His promise. And a thousand years are as one day - That is, no delay is long to God. Therefore 'He is longsuffering:' He gives us space for repentance, without any inconvenience to Himself. In a word, with God time passes neither slower nor swifter than is suitable to Him and His economy; nor can there be any reason why it should be necessary for Him either to delay or hasten the end of all things. How can we comprehend this?"[18]

(3) Following the extinguishing of the heavenly luminaries, then shall appear the "sign of the Son of man" in the completely darkened heaven. Wesley explained the timing of this sign's advent - "It

seems, a little before He Himself descends." The nature of the sign - "perhaps the cross."[19] Then, Christ comes in glory, sending his attending angels throughout the earth to "gather together his elect" - that is, "all that have endured to the end in 'the faith which worketh by love.'"[20]

Concerning the time when these "two grand events" occur, according to Matthew's gospel, Wesley observed - "Concerning the time of the destruction of Jerusalem, He answers, verse 34. Concerning the time of the end of the world, He answers, verse 36."[21] Verse 34 reads - "This generation shall not pass away till all things be done." Verse 36 reads - "But of that day and hour knoweth no man, neither the angels of heaven, but my Father only." So, the generation that heard Jesus predict the destruction of the holy city and its temple witnessed the fulfillment thirty-nine to forty years later in A.D. 70. But, as for his prediction of the coming of the Son of man and the end of the world, many generations have passed since that was foretold by Christ. During Christ's earthly life, "no man" knew when the Parousia would be. However, Wesley believed that the exact time was revealed later to "St. John" in the APOCALYPSE,[22] making a rather exhaustive study of that book necessary. But, we will not undertake such a treatment until later in our study.

The passage from Matthew concludes with a description of faithful servants and faithless servants at the Parousia. The word "servant" in the Greek text of Matthew is *doulos*, meaning a *slave*. Wesley observed that at Christ's coming some persons would be "taken" into God's "immediate protection" while others would be "left... to share the common calamities."[23] Nowhere does Wesley refer to those being "taken" as being "raptured." While he often used the term in an aesthetic sense, he never used it eschatologically. The popular belief in the "Rapture" is a Post-Wesleyan invention of the mid-nineteenth century. "Who then is the faithful and wise servant?" Wesley answered -

"Which of you aspires after his character? Wise - every moment retaining the clearest conviction that all he now has is only entrusted to him as a steward.

Faithful - Thinking, speaking, and acting continually in a manner suitable to that conviction."[24]

The way of the faithless servant is hard because he is actually one who has "put away faith and a good conscience." So, at the *Parousia*, he is assigned an eternal portion with "the hypocrites" -

"(Such a person is) the worst of sinners, as upright and sincere as he was once. If ministers are the persons here primarily intended, there is a peculiar propriety in the expression. For no hypocrisy can be baser than to call ourselves ministers of Christ, while we are slaves of avarice, ambition, or sensuality. Where such are found, may God reform them by His grace, or disarm them of that power and influence which they continually abuse to His dishonour, and to their own aggravated damnation."[25]

Wesley maintained that Christ's coming begins an immediate period of judgment as taught by Jesus in the Gospel of Matthew, in which the faithful are divinely protected and the faithless are subjected to damning calamities as a consequence of their judgment. He found additional support for this position in St. Paul's First Epistle to the Thessalonians IV, 16ff -

"For the Lord himself shall descend with a shout, with the voice of an archangel, and with the trumpet of God; and the dead in Christ shall rise first: Then we who are alive who are left shall be caught up together with them in the clouds, to meet the Lord in the air."

Such a gathering up into the clouds, for Wesley, is not for the purpose of God's elect escaping tribulation on earth and being granted heavenly rest until a future judgment - as in the Rapture theory. Never! At Christ's coming, he raises the saints to his side in the clouds to help him begin judging those remaining on earth - "The wicked will remain beneath, while the righteous, being absolved, shall be assessors with their Lord in the judgment...in heaven (the lower heaven)."[26]

Closely connected with this Pauline passage is II Thessalonians II, 1ff - which Wesley acknowledged as a clarification by Paul of his teaching concerning the time of the Parousia in his first letter to Thessalonica. The text reads as follows -

"Now I beseech you, brethren, concerning the appearing of our Lord Jesus Christ, and our gathering together unto him. That ye be not soon shaken in mind, or terrified, neither by spirit, nor by word, nor by letter as from us, as if the day of the Lord were at hand. Let no man deceive you by any means, for that day shall not come, unless the falling away come first, and the man of sin, be revealed, the son of perdition; Who opposeth and exalteth himself above all that is called God, or that is worshipped; so that he sitteth in the temple of God as God, declaring himself that he is God."

Wesley saw this sign of falling away before the coming of Christ as an expansion of Christ's teaching in Matthew. However, the Pauline version adds the person of "the man of sin" or "the son of perdition." Wesley's identification of this person in the *EXPLANATORY NOTES* is important to all that follows -

". . .the falling away - From the pure faith of the gospel. . .this began even in the apostolic age. . .The man of sin, the son of perdition - Eminently so called, is not come yet (in Paul's time). However, in many respects, the Pope has an indisputable claim to these titles. He is, in the emphatical sense, the man of sin, as he increases all manner of sin above measure. And he is, too, properly styled, the son of perdition, as he has caused the death of numberless multitudes, both of his opposers and followers, destroyed innumerable souls, and will himself perish everlastingly. He it is that opposeth himself to the emperor, once his rightful sovereign (Henry IV of Germany in the eleventh century): and that exalteth himself above all that is called God, or that is worshipped - Commanding angels and putting kings under his feet, both of whom are called gods in Scripture; claiming the highest power, the highest honour; suffering himself, not once only, to be styled God or vice-god. Indeed, no less is implied in his ordinary title, 'Most Holy Lord,' or 'Most Holy Father.' So that he sitteth enthroned...in the temple of God, Mentioned in Rev. XI, 1. . .Declaring himself that he is God - Claiming the prerogatives which belong to God alone."[27]

The man of sin, or son of perdition, is inextricably linked to what Wesley termed "the deep, secret power of iniquity," which is the opposite of the power of godliness, and now in the world. Such power was at work, however, long before the man of sin was revealed. It is, though, forever his trademark in all ages. It is characterized by love of "honour" and desire of "power." It seeks the "entire subversion of the Gospel of Christ." Paul called this phenomenon "the mystery of iniquity," and it works not only in the Roman Church but in all others as well.[28] Some of its achievements are - (1) human inventions added to the scriptures; (2) external works put in the place of faith and love; and (3) other mediators (mediatrix?) besides Christ Jesus. Historically, the man of sin has been restrained, not by the Holy Spirit as some claim, but by various emperors, Christian and pagan, Goths, Lombards, and Carolingians.[29] When every prince is taken away, then the man of sin will be "consumed" by Christ's immediate power at "the very first appearance of His glory."[30]

In addition to exegeting this passage from II Thessalonians, Wesley often preached upon it. His first recorded sermon on the text (II, 7 - "The mystery of iniquity doth already work") was preached at the Foundery in London, on October 9, 1754. After that occasion, Wesley preached the sermon regularly in other societies. The sermon development traces the biblical tension between the mystery of godliness and the mystery of iniquity from Adam onward, even within the family of Abraham. But, "in the fulness of time, when iniquity of every kind. . . had spread over all nations, and covered the earth like a flood, it pleased God to lift up a standard against it by 'bringing his first-begotten into the world.'"[31] The Church that Christ established on the day of Pentecost enjoyed the mystery of godliness as evidenced by their "having all things in common." Then, tragically, "the mystery of iniquity" began to work again and obscured "the glorious prospect!"[32] Ananias and Sapphira (Acts V, 1ff) gave place to the devil, and lied to the Holy Spirit, making "shipwreck of faith and a good conscience," drawing "back to perdition."[33] The plague of iniquity was immediately "stayed in the first Christian Church, by instantly cutting off the infected persons" (Acts V, 11), and great fear came upon the whole Church. The mystery of iniquity revived later in that Church when partiality was shown some widows over others at the daily food distribution center (Acts VI, 1ff), and a "root of bitterness" sprang up and many "were defiled."[34] The circumcision controversy next divided the Church, then Paul and Barnabas parted company because Barnabas had "a fit of anger."[35] The mystery of iniquity also worked in the Church at Rome (Romans XVI, 17) and at Corinth (I Corinthians).[36] St. James attacked that "grand pest of Christianity, a faith without works," clearly a by-product of the mystery of iniquity.[37] St. Peter warned against the growing of tares among the wheat - Christians who "walk after the flesh. . . in the lust of uncleanness, like brute beasts, made to be taken and destroyed."[38] "Such is the authentic account of 'the mystery of iniquity' working even in the apostolic Churches. . . given by the Apostles themselves."[39] But, after the death of the apostles, the mystery of iniquity experienced phenomenal growth within the Church. Tertullian (A.D. 160-240) reported that Christian tempers, lives, and manners were no different than those of their heathen neighbours.[40] Bishop Cyprian of Carthage (died, A.D. 258) described the African Church as being "immersed in ambition, envy, covetousness, luxury, and all other vices." Wesley added, "The same as the Christians of England are now."[41] The mystery of iniquity, however, made its greatest advance when Emperor Constantine "called himself a Christian, and poured in a flood of riches, honours, and power, upon the Christians; more especially upon the Clergy."[42] The Christians then ran headlong "into all manner of vices," and iniquity was everywhere visible for the entire world to behold. Many interpreters of this sad epoch have declared the rise of state-religion under Constantine to be the New Jerusalem on earth. Wesley rejected such a preposterous identification.[43] From Constantine to the Protestant Reformation, Wesley argued, there was no truly Christian nation or city. While the Reformation rediscovered something of Apostolic Christianity, "it was not carried far enough!"[44] "Now, let any one survey the state of Christianity in the Reformed parts of Switzerland; in Germany, or France; in Sweden, Denmark, Holland; in Great Britain and Ireland." How little are any of "these Reformed Christians better than heathen nations! . . . have they more justice, mercy, or truth, than the inhabitants of China, or Indostan? O no! we must acknowledge with sorrow and shame, that we are far beneath them!"[45] Wesley continued his analysis - Wherever Christianity has spread, the apostasy from God has spread also. Ironically, the Christian religion was raised up by God for the "healing of the nations," for the "saving from sin by means of the Second Adam." But, it does not answer this end - "It never did; unless for a short time at Jerusalem."[46] What the world sees when it observes the life of a "Christian" is not necessarily that of a Christian -

"Of Christians, do you say? I doubt whether you ever knew a Christian in your life. When Tomo Chachi, the Indian Chief, keenly replied to those who spoke to him of being a Christian, 'Why, these are Christians at Savannah! These are Christians at Frederica!' - the proper answer was, 'No, they are not; they are no more Christians than you and Sinauky.' But are not these Christians in Canterbury, in London, in Westminster? No; no more than they are angels. None are Christians, but they that have the mind which was in Christ, and walk as he walked. 'Why, if these only are Christians,' said an eminent wit, 'I never saw a Christian yet.' I believe it; You never did; and, perhaps you never will; for you will never find them in the grand or gay world. The few Christians that are upon the earth, are only to be found where you never look for them."[47]

THE SIGNS OF THE END IN THE APOCALYPSE

The words of I John II, 18, with Wesley's exegesis of them, make a dynamic prologue for the signs in the book of Revelation -

"Little children, it is the last time: and as ye have heard that antichrist cometh, so even now there are many antichrists; whereby we know that it is the last time."

"My little children, it is the last time - The last dispensation of grace, that which is to continue to the end of time, is begun. *Ye have heard that antichrist cometh* - Under the term antichrist, or the spirit of antichrist, he includes all false teachers, and enemies to the truth; yea, whatever doctrines of men are contrary to Christ. It seems to have been long after this that the name antichrist was appropriated to that grand adversary of Christ, 'the man of sin' (2 Thess. II, 3). Antichrist, in St. John's sense, that is, antichristianism

spreading from his time till now; and will do so, till that great adversary arises, and is destroyed by Christ's coming."[48]

In his preliminary remarks to the commentary on Revelation, Wesley noted that the universal experience of serious readers of the book involves feeling "their hearts extremely affected" by its first and last parts. However, the "intermediate" parts of the prophecy leave most readers baffled. Wesley had long been among these readers -

". . . the intermediate parts I did not study for many years, as utterly despairing of understanding them, after the fruitless attempts of so many wise and good men: and perhaps I should have lived and died in this sentiment, had I not seen the works of the great Bengelius. But these revived my hopes of understanding even the prophecies of this book; at least many of them in some good degree: for perhaps some will not be opened but in eternity. Let us, however, bless God for the measure of light we may enjoy, and improve it to His glory."[49]

While Wesley's *EXPLANATORY NOTES* on the Apocalypse were essentially drawn from Bengelius's *GNOMON NOVI TESTAMENTI* and his *EKLARIE OFFENBARUNG*, no one should suppose that Wesley's heart was absent from his use of such material. He was thoroughly convinced that Bengelius had decoded this mysterious book, and he was committed to enlightening his Methodist pilgrims with this ingenious system of interpretation. Consequently, Wesley translated the German originals into English, abridging as he moved along, altering arguments and adding arguments of his own where Bengelius was "not full."[50] Wesley added a disclaimer -

"Yet I by no means pretend to understand or explain all that is contained in this mysterious book. I only offer what help I can to the serious inquirer, and shall rejoice if any be moved thereby more carefully to read and more deeply to consider the words of this prophecy."[51]

The true author of the Apocalypse, according to Wesley, was not John but Jesus Christ. John the Presbyter only moved the pen. The recipients were addressed as Christ's "servants" or "slaves." The message of Christ to his *douloi* (servants/slaves) concerned "things to come, many. . .near, intermediate, remote; the greatest, the least; terrible, comfortable; old, new; long, short; and these interwoven together, opposite, composite; relative to each other at a small, at a great, distance."[52] These were the things that his servants in the seven churches of Asia wanted and needed to know. While the book was intended primarily for the seven churches of Asia, it also speaks to the servants of Christ in other nations and ages.

The prophecies of the book began to be fulfilled shortly after it was written, Wesley reasoned. Accordingly, the book, rightly understood, reveals these things in various stages, the earliest of which began with the first circulation of the book.[53] Unfortunately, he continued, some scholars and many nonscholars have "miserably handled this book," for which reason, many Christian pilgrims are afraid to "touch" this prophecy.[54] However, there are some excellent literary resources which can help the serious seeker of knowledge in these matters - Bengelius's works for instance, and, of course, *THE EXPLANATORY NOTES UPON THE NEW TESTAMENT.* These works, Wesley maintained, rightly identify the correct meaning of the book's symbols - seven epistles, seven seals, seven trumpets, seven phials - each of these sevens being divided by four and three - then, seven stars, seven candlesticks, the Lamb, His seven horns and seven eyes, the incense, the dragon, the heads and horns of the beasts, the fine linen, the testimony of Jesus, and many others.[55] Moreover, many of these prophetic images have their origin in the Old Testament prophets. Daniel furnished Revelation with the "description of the man child." Isaiah furnished "the promises of Sion" (Zion). Jeremiah gave the "judgment of Babylon." "The architect of the holy city" was Ezekiel. The emblems of the horses and the candlesticks were first advanced by Zechariah. This last prophet also spoke of two olive trees, as did St. John but with a different meaning.[56]

Before his death, Christ spoke of things to come in a short while, adding a very brief description of the last things. In the book of Revelation, however, Christ revealed in some detail the intermediate things to come, things between the destruction of Jerusalem and the Temple and the last things. Consequently, this book describes special events treated nowhere else in Scripture, Wesley explained.[57] Following Bengelius, Wesley placed "the intermediate things" in a special chapter sequence, related to the vantage point of the eighteenth century - chapters VI - IX, things already fulfilled; chapter X - XIV, things presently being fulfilled; and chapters XV - XIX, things to be fulfilled shortly. As for chapters I-III, they form the introduction to the entire book. Chapters IV and V represent the "proposition" - the Lamb breaking the seals which keep the prophecy closed. Chapters XX - XXII treat things to come at a greater distance, truly the last things.[58]

A. It would be unpardonable to gloss over Wesley's treatment of the first three chapters of the Revelation, what he termed the "Introduction." His observations on key words, phrases, and concepts need analysis if any meaningful understanding of his eschatology is to be had. (1) Chapter I - Verse 4 indicates that the seven churches to whom St. John wrote were those that had received his greatest labors in Asia. There were many believing Jews in these churches, as well as believing Gentiles. Hence, the reference to God as the one "who is, and who was, and who cometh, or, who is to come," was especially appreciated by the Jewish Christians in these churches because it was "a wonderful translation of the great name JEHOVAH."[59] Also mentioned in this verse are the "seven spirits that are before his throne." Wesley identified these spirits as "the seven lamps which burn before his throne," that is, the Holy Ghost of God. Wesley observed that "seven was a sacred number in the Jewish church: but it did not always imply a precise number. It sometimes is to be taken figuratively, to denote completeness or perfection."[60] In no wise does the phrase, "the seven spirits," weaken the traditional doctrine of the Trinity. Wesley claimed the phrase describes his operations and not his essence.[61] In verse 5, Christ is acclaimed as "the faithful witness, the first begotten from the dead, and the prince of the kings of the earth." Wesley saw Christ as the perfect witness to God's will before his death, as the "firstfruits" of them that slept in death, and as the possessor "of all power both in heaven and earth." These titles shall be exchanged for another at his coming - "King of kings, and Lord of lords."[62] "Behold, he cometh," verse 7, refers to Christ's glorious coming, Wesley explained, "the preparation for this began at the destruction of Jerusalem, and more particularly at the time of writing this book; and goes on, without any interruption, till that grand event is accomplished. . . It is never said. . . He *will* come; but *He cometh*." Wesley was adamant - "It is not said, He cometh *again*: For when He came before, it was not like Himself, but in the form of a servant. But His appearing in glory is properly His coming; namely, in a manner worthy of the Son of God."[63] The Alpha and Omega theme of verse 8 undergirds the entire book of Revelation for Wesley -

> "The Lord God is both, *the Alpha*, or beginning, and *the Omega*, or end, of all things. God is the beginning, as He is the Author and Creator of all things, and as He proposes, declares, and promises so great things: He is the end, as He brings all the things which are here revealed to a complete and glorious conclusion. Again, the beginning and end of a thing is in Scripture styled the whole thing."[64]

In his exposition of verse 9, Wesley dated the writing of the Apocalypse in the reigns of the Roman emperors Domitian (A.D. 81-96) and Nerva (A.D. 96-98). The student of history will remember the persecution of Christians throughout the Empire which followed Domitian's claim to divine honors. Wesley saw John the Presbyter as "a banished man" of this era.[65] Moreover, the prophet John received this vision on the "first day of the week" or the Lord's Day, while banished on Patmos. Wesley noted that the early Church expected Christ, who was raised from the dead on the first Lord's Day, to come in glory and judgment on that same day of the final week.[66] Verses 10-20 treat John's vision of the heavenly Son of Man, who once was dead but now is alive forever, possessing the keys of death and of hell. While Wesley's entire exegesis of this vision is worthy of citation, only a few observations are useful to this study. Wesley's understanding of the "keys of death and of hell" is crucial -

> "The keys of death and of hades - That is, the invisible world. In the interme-
> diate state, the body abides in death, the soul in hades (Abraham's Bosom).
> Christ hath the keys of, that is, the power over, both; killing or quickening of
> the body, and disposing of the soul, as it pleaseth Him. He gave St. Peter the
> keys of the kingdom of heaven; but not the keys of death or of hades. How
> comes then his supposed successor at Rome by the keys of purgatory?"[67]

The "mystery of the seven stars" of verse 20 is quickly and deci-sively decoded by Wesley. The seven stars in the right hand of the Son of Man, called the "angels of the seven churches," are the "pastors, bishops, and overseers" of these churches, and, in the larger sense, all "ruling ministers" of Christ's Church.[68] It should not be overlooked that Wesley's view of the clergy was a High Church position, depending on a belief in apostolic lineage of doc-trine, ethics, and liturgy.

(2) Chapters II and III - Wesley regarded the seven letters of these chapters to be "a kind of sevenfold preface to the book." He further analyzed them in terms of common elements -

> "There is in each of these letters,
> 1. A command to write to the angel of that church;
> 2. A glorious title of Christ;
> 3. An address to the angel of that church, containing
>   A testimony of his mixed, or good, or bad state;
>   An exhortation to repentance or steadfastness;
>   A declaration of what will be; generally, of the Lord's coming;
> 4. A promise to him that overcometh, together with the exhortation,
>   'He that hath an ear to hear, let him hear.'"[69]

The address in each letter is in plain words, the promise is always in symbolic or figurative words. "In the address our Lord speaks to the angel of each church which then was, and to the members thereof directly." Wesley did not hold the view that each of the seven churches represents an era of church history. Such an interpretation was explicitly rejected by him as the product of imagination.[70] However, Wesley claimed, the promises are for Christians of every age, should they overcome through faithfulness to Christ.[71] John depicted Christ as standing in the midst of the seven golden candlesticks which represented the seven churches of Asia. Christ spoke to each church in turn, to Ephesus, Smyrna, Pergamos, Thyatira, Sardis, Philadelphia, and Laodicea. Wesley observed the following important elements in each message - (a) Ephesus - Christ "knew" the labors of pastor and laity, their patience, their rejection of evil men, their trial of impostors, and their faithfulness to the name of Christ. But, Christ chided them for their loss of agape (perfect love) which they had at first. They are commanded to remember, repent, and do the first works. Yet Christ commended them for "hating the works of the Nicolaitans, which I hate." Wesley said of this sect here condemned -

"Probably so called from Nicolas, one of the seven deacons (Acts VI, 5). Their doctrines and lives were equally corrupt. They allowed the most abominable lewdness and adulteries, as well as sacrificing to idols, all which they placed among things indifferent, and pleaded for as branches of Christian liberty."[72]

(b) Smyrna - Christ claimed knowledge of the affliction and poverty of this church, but pronounced that it was rich spiritually. False Jews, of the synagogue of Satan, caused the persecution of this church, but such affliction was to last for only ten days - Wesley took the "ten days" to mean until "the end of Domitian's persecution."[73] (c) Pergamos - This church lived in a city given over completely to idolatry. Hence the city is the "throne of Satan." Yet the church had not denied the faith, even though Antipas - a Christian martyr according to Wesley - was slain under Domitian's savagery. However, the church was unduly permissive and allowed the Nicolaitans to remain within its fellowship.[74] (d) Thyatira - Here the church consisted of a little flock which was dominated by a "Jezebel" - Wesley claimed this woman was "the wife of the pastor" as many ancient authorities held. Usurping the ministry of her husband, she taught the little flock to fornicate, both literally and symbolically, emphasizing the "deep things of Satan."[75] (e) Sardis - The church of Sardis had a reputation for being alive, but Christ condemned it for being dead. Nevertheless, there were "a few" in the church who were still alive to God, no thanks to the angel or pastor![76] (f) Philadelphia - This church was exemplary, an almost ideal church, not needing any chiding or condemnation from Christ. Consequently, Christ promised to keep this church from "the hour of temptation" - Wesley identified this "hour" as the "time of persecution under the seemingly virtuous emperor Trajan." He added -

> "The two preceding persecutions were under those monsters Nero and Domitian; but Trajan was so admired for his goodness, and his persecution was of such a nature, that it was a temptation indeed, and did thoroughly try them that dwelt upon the earth."[77]

It must be noted that Wesley did not accept the notion that the "hour of temptation" was the "Great Tribulation" of twentieth century premillenialism. (g) Laodicea - Spiritually the church of this city was "lukewarm" - only partially committed to Christ. As lukewarm water induces vomiting, so such lukewarm faith induces Christ to "spew" such persons from his mouth. Spiritual lukewarmness is synonymous with being spiritually poor, blind, and naked, but Christ is willing to be the great physician and heal this condition where he is invited in.[78]

Before leaving the subject of the letters to the seven churches, the stress of Wesley on Christ's promises to them needs brief attention. Wesley's own words best summarize the case -

> "In these seven letters twelve promises are contained, which are an extract of all the promises of God. Some of them are not expressly mentioned again in this book as 'the hidden manna,' the inscription of 'the name of the new Jerusalem,' the 'sitting upon the throne.' Some resemble what is afterwards mentioned, as 'the hidden name' (Rev. XIX, 12); 'the ruling the nations' (Rev. XIX, 15); 'the morning star' (Rev. XXII, 16); And some are expressly mentioned, as 'the tree of life' (Rev. XXII, 2); freedom for 'the second death' (Rev. XX, 6); the name in 'the book of life' (Rev. XX, 12; XXI, 27); the re-

maining in 'the temple of God' (Rev. VII, 15); the inscription of 'the name of God and of the Lamb' (Rev. XIV, 1; XXII, 4). In these promises sometimes the enjoyment of the highest good, sometimes deliverance from the greatest evils, is mentioned. And each implies the other, so that when either part is expressed, the whole is to be understood. That part is expressed which has most resemblance to the virtues or works of him that was spoken to in the letter preceding."[79]

B. With Chapter IV and V of the Apocalypse, Wesley argued, the main prophecy of the book is introduced as a sealed scroll, containing within and without the entire program by which God will consummate all things through Christ. It is not until chapter VI, however, that Christ begins to break the seals so the scroll may be opened - as a set of blue-prints - and the eschatological program may proceed according to the plan. The fourth and fifth chapters, Wesley explained, must be viewed together, showing the divine origin and authority of the prophetic plan.

The prophet John heard a loud command - "Come up hither, and I will show thee things which must be hereafter" (IV, 1). Wesley observed that the prophet was not carried up bodily into heaven, as some interpreters have attempted to argue, but, rather, he was lifted up "in spirit; which was immediately done."[80] Suddenly, the prophet was standing in the heavenly court, somewhat reminiscent of the prophet Isaiah's spiritual transport into the heavenly temple (Isaiah VI, 1ff). God - as King, Governor, and Judge - sat enthroned in "the centre" of heaven, "with a visible lustre, like that of sparkling precious stones," like jasper - "a symbol of God's purity." Another symbol of God's appearance was "sardine stone" which Wesley claimed was "as blood-red colour... an emblem of His justice, and the vengeance He was about to execute on His enemies."[81] A rainbow was around the throne, symbolizing God's everlasting covenant. Also, surrounding God's throne were twenty-four thrones, with twenty-four elders seated upon them. Wesley believed these elders were saints from the former age, before Christ's time, representing "the whole body of the saints." Generally they remain seated, but in worship they fall down prostrate before God. They are clothed in white, and their golden crowns show that "they had already finished their course and taken their place among the citizens of heaven. They are never termed souls, and hence it is probable that they had glorified bodies already (Matt. XXVII, 52)."[82] From the throne of God issued lightning flashes, voices, and thunders. Before this throne was "a sea as of glass, like crystal, wide and deep, pure and clear, transparent and still." Above the divine throne - towards the four quarters of east, west, north, and south - were four living creatures, not *beasts*. Wesley thought these creatures were the cherubim mentioned by Isaiah and Ezekiel. The first creature was like a lion, signifying undaunted courage. The second like a calf, signified unwearied patience. The third, with the face of a man, signified prudence and compassion. The fourth like an eagle, signified activity and vigor.[83] Each had six wings and were full of eyes. They behold all of the created order and recognize the divine plan at work within it, and they exclaim, "Holy, holy, holy, is the Lord God, Almighty, who was, and who is, and who cometh." The elders before the heavenly throne, upon hearing the chanting of these four creatures, fall prostrate before God and respond saying, "Worthy art Thou, O Lord our God, to receive the glory, and the honour, and the power: for Thou hast created all things, and through Thy will they were and are created."[84]

Chapter V is a continuation of John's vision of the heavenly court. God's right hand, Wesley argued, signifies "His all-ruling power." In that right hand God held a book, written within and without, bound shut with seven seals. The book is figurative and not literal -

"It is scarce needful to observe, that there is not in heaven any real book of parchment or paper, or that Christ does not really stand there, in the shape of a lion or of a lamb. Neither is there on earth any monstrous beast with seven heads and ten horns. But as there is upon earth something which, in its kind, answers such a representation, so there are in heaven divine counsels and transactions answerable to these figurative expressions. All this was represented to St. John at Patmos, in one day, by way of vision."[85]

Wesley explained that the "book" was more of a scroll. Although there are seven seals which keep it bound shut, the scroll represents "all power in heaven and earth given to Christ." It is a divine plan, idea, pattern, or blue-print of God's program of finalizing his will in Jesus Christ. "A copy of this book is contained in the following chapters," Wesley reported.[86] Its contents include the breaking of the seven seals by the Lion who is also a Lamb, seven trumpets as part of the breaking of the seventh seal, the kingdom of the world shaken so it may become the Kingdom of Christ, seven phials of God's wrath as part of the seventh trumpet, and the power of the beast being broken.[87] Christ, the Lion and the Lamb, received the sealed book from God, as the four creatures and the twenty-four elders fell down and sang a new song - "Worthy art thou to take the book, and to open the seals thereof: for thou wast slain, and hath redeemed us to God by thy blood. . . and hast made (us) unto our God kings and priests: and (we) shall reign over the earth.[88] Now Christ was ready to begin breaking the seven seals.

C. In the *EXPLANATORY NOTES UPON THE NEW TESTAMENT*, Wesley treated chapters VI through IX of Revelation as signs of Christ's coming which have already taken place in history. The earliest of these signs began just after the time the letters to the seven churches of Asia were received. They were associated with the vision of the breaking by Christ of the first four seals. These four signs were historically interconnected, Wesley believed, and "relate to visible things, toward the four quarters to which the four living creatures look."[89] Before exegeting any further, Wesley sounded some reminders - There is no way to explain everything in this book. Revelation relates to the whole world wherein Christians are diffused. Our interpretation must not judge one thing as great and another small, for God sees things differently than does man. And we must take into consideration what has already been fulfilled, not describing "as fulfilled what is still to come."[90]

The breaking of the first four seals by Christ have already been fulfilled in history. In each seal passage, a different horseman appears. To rightly understand each passage, one must consider first the horseman himself and then what he does. Each rider is "an emblematical prosopopoeia" representing a swift power - raised up by God - "bringing with it either, (1) a flourishing state; or (2) bloodshed; or (3) scarcity of provisions; or (4) public calamities." Wesley's general description is worth noting -

> "With the quality of each of these riders the colour of his horse agrees. The fourth horseman is expressly termed 'Death;' the first, with his bow and crown, 'a conqueror;' the second, with his great sword, is a warrior, or, as the Romans termed him Mars; The third, with the scales, has power over the produce of the land.
>
> The action of every horseman intimates further, (1) toward the east, widespread empire, and victory upon victory; (2) toward the west, much bloodshed; (3) toward the south, scarcity of provisions; (4) toward the north, the plague and various calamities."[91]

The Wesleyan exposition of these four riders, as signs of Christ's coming already fulfilled, deserves our attention. (1) The rider with a bow and a crown, astride a white mount, "betokens victory, triumph, prosperity, enlargement of empire, and dominion over many people." The historical key to unlocking the mystery of this rider is Nerva, the successor to Domitian as emperor of Rome. Nerva ruled from A.D. 96 to 98. Just before his death, Nerva appointed Trajan to be his successor. "Trajan's accession to the empire seems to be the dawning of the seven seals," Wesley affirmed. God gave the emperorship to Trajan, by the hand of Nerva, since by family descent he never could have attained it. In the year A.D. 108, Trajan went on a military conquest eastward from Rome - conquering Armenia, Assyria, Mesopotamia, and countries beyond the Tigris River, Trajan had no other purpose except to conquer, as it was prophesied of the fourth emperor in Daniel II, 40 and VII, 23, "that he should devour, tread down, and break in pieces the whole earth."[92] (2) The second rider appears with a great sword in hand, seated upon a red horse, intent upon taking "peace from the earth." Red, figuratively, indicates bloodshed. Wesley noted that, in A.D. 75, Emperor Vespasian (first of the Flavian Emperors) erected a temple to peace in Rome but since that time onward there was no peace in the Roman world, especially in the western part. Trajan was largely responsible for this, as was Decebalus, king of Dacia in the west, who waged a five year war against Rome, accounting for the greatest bloodshed in Roman history.[93]   (3) The third rider rode southward on a black horse - "a fit emblem of mourning and distress; particularly of black famine, as ancient poets term it." The rider carried a pair of scales in his hand, the scales signifying scarcity of food, implying that nature interrupts its productivity throughout the yearly growing cycles. The hand of Christ is in this, bringing victory in nature as well as in war. Wesley observed the cost of a measure of wheat during this time was a "penny." In the Greek text of Revelation, however, it is *denariou* (genitive, singular), about twenty cents in present exchange. Wesley was correct in noting that one such coin's purchasing power for wheat in this passage could supply only a slave's daily allowance. Such short supply of wheat and barley, inflating food prices, occurred "in Egypt under Trajan." Oil and wine, however, were not affected. "In Egypt, which lay south of Patmos. . . which used to be the granary of the empire, there was an uncommon dearth at the very beginning of his reign; so that he was obliged to supply Egypt itself with corn from other countries," Wesley claimed. In the thirteenth year of Trajan's administration, Egypt was again struck with scarcity when the Nile did not flood according to its annual habit. In fact, Wesley added, the parts of Africa bordering the Nile suffered as badly as Egypt.[94] And (4) the fourth rider, on a pale horse, rode "toward the north." This rider was given the name "Death," and Hades followed after him to receive whomever Death might slay. Death had power over the fourth part of the earth, that is, over the north. Death kills by the scimitar (a short, curved, single-edged sword used by the ancient Persians), by famine, by plague, and by wild beasts. These forms of death overtook "the fourth part of men" within the Roman Empire from the time of Trajan onward, Wesley argued. He quoted Aurelius Victor - "At that time the Tyber overflowed much more fatally than under Nerva, with a great destruction of houses; and there was a dreadful earthquake through many provinces, and a terrible plague and famine, and many places consumed by fire."[95] As for wild beasts attacking humans, Wesley maintained that during Trajan's time the phenomenon increased alarmingly, as if "an uncommon fierceness and strength" had been given them by God. Then too, "war brings on scarcity, and scarcity pestilence, through want of wholesome sustenance; and pestilence by depopulating the country, leaves the few survivors an easier prey to the wild beasts."[96] So, the four horsemen come forth unto the earth, as Christ has planned, "subsisting by His power, and serving His will, against the wicked, and in defence of the righteous."[97]

These first four seals, broken by Christ, relate to the visible world. The fifth, sixth, and seventh seals, however, relate to the invisible world of spirits - "the fifth, to the happy dead, particularly the martyrs; the sixth, to the unhappy; the seventh, to the angels, especially those to whom the trumpets are given."[98] The background of these last three seals, as Wesley understood it, is the Church warring under its general, Jesus Christ, and "the world warring under Satan." The invisible hosts of both Christ and Satan are included in this spiritual war. Now when Christ broke the fifth seal, the Elder John saw the souls of the martyrs, who had been slain for the Word of God, lying at the foot of one of the heavenly altars. Wesley identified two different altars in the Apocalypse - the "golden altar of incense" in IX, 13, and the "altar of burnt-offerings" here and in VIII, 5; XIV, 18; and XVI, 7. Resting prostrate at the altar of burnt-offerings - having been offered up to God through the fire of persecution - these holy souls await the time when their "blood shall be avenged upon Babylon" but "not yet."[99] While resting, these souls cry out with a loud voice - "How long, O Lord, thou Holy One and True, dost Thou not judge and avenge our blood on them that dwell upon the earth?" Wesley claimed that this question "did not begin now, but under the first Roman persecution." To divert their minds from the persistent question of vengeance, God gave each of these martyred souls a white robe, symbolizing innocence, joy, and victory, "in token of honour and acceptance."[100] They were ordered to rest for a "time" (*chronon mikron* in the Greek text). Wesley saw great importance to this concept of time -

"A time - This word has a peculiar meaning in this book, to denote which we may retain the original word chronos. Here are two classes of martyrs specified: the former killed under heathen Rome; the latter, under papal Rome. The former are commanded to rest till the latter are added to them. There were many of the former in the days of John: the firstfruits of the latter died in the thirteenth century. Now, *a time*, or *chronos* is 1111 years. This chronos began A.D. 98, and continued to the year 1209, or from Trajan's persecution to the first crusade against the Waldenses."[101]

The sixth seal was broken by Christ. Immediately there was a great earthquake, the sun was darkened, the moon turned blood-red, stars fell to earth, the first heaven departed, every mountain and island were moved from their place, all the human inhabitants of the earth hid in caves and prayed to the mountains and rocks to fall on them in order to hide them from "the wrath of the Lamb: For the great day of his wrath is come; and who is able to stand?"[102] Wesley's commentary on this passage breaks with the historical-identification-treatment he earlier established for this part of the Apocalypse. In the *EXPLANATORY NOTES*, he does not name Nerva, Trajan, or one pope in relation to this sixth seal. He hastily stated -

"This sixth seal seems particularly to point out God's judgment on the wicked departed. St. John saw how the end of the world was even then set before the unhappy spirits. This representation might be made to them, without anything of it being perceived on the earth."[103]

In other words, the sixth seal is a parable about the future end of the world, addressed to wicked spirits of angels and men alike, warning them of what lies ahead, "without anything of it" - earthquake, darkened sun, bloodied moon, etc. - "being perceived on the earth." It is a "flash-ahead" to what will be.

The last seal broken by Christ (Revelation VI, 1ff) was of great significance in Wesley's eschatology. The passage opens with mention of five angels. Wesley spoke of the first four as being "evil."[104] He did not give any reasons for this identification. However, every Greek New Testament student, perusing the passage in the original, knows why - These four angels are commanded to "hurt not the earth" (vs. 3). The Greek word translated "hurt" is *adikesai*, which implies an "unjust act" more than "hurt." Wesley continued to extol their evil intentions - "These four angels would willingly have brought on all the calamities that follow without delay. But they were restrained till the servants of God were sealed, and till the seven angels were ready to sound."[105] Each of these spirits stood on a corner of the earth - east, west, south, north, keeping the four winds from blowing on the earth - "which else might have softened the fiery heat."[106] The fifth angel, identified by Wesley as being good, then ascended from the east, possessing the seal of the living and true God. It was he who restrained the four evil angels "to whom it was given to hurt the earth and the sea."[107] Other good angels joined this fifth angel in placing the seal of God on the foreheads of the twelve tribes of Israel, securing them to God despite the "impending calamities. . . whereby they shall be as clearly distinguished from the rest, as if they were visibly marked on their foreheads."[108] While some interpreters of Wesley have viewed him in antiSemitic terms, his commentary here takes a sympathetic turn toward historic Israel.

It has been common for Christian exegetes to treat the Israel of Revelation VII as the "New Israel" or the New Testament Church, since, they argue, the Old Israel has been replaced because of its unbelief in Jesus as the Messiah. Hence, their interpretation of the "sealing of Israel" emphasizes the Church being mystically divided into twelve spiritual tribes, from which twelve thousand persons per tribe are sealed. Wesley would have none of this! He saw the Church of Jesus Christ as existing in Eden, immediately after the Fall, and continuing through history, including Abraham, Moses, *et. al*, into the Gospel dispensation. Like St. Paul (Romans IX), Wesley could not believe that Israel's unbelief was permanent and that God had cast off his covenant people. Consequently, Wesley exegeted the sealing of Israel in Revelation VII accordingly -

"*Of the children of Israel* - To these will afterwards be joined a multitude out of all nations. But it may be observed, this is not the number of all the Israelites who are saved from Abraham or Moses to the end of all things; but only of those who are secured from the plagues which were then ready to fall on the earth. It seems as if this book had, in many places, a special view to the people of Israel."[109]

The one hundred and forty-four thousand sealed from the twelve tribes of Israel, together with a great innumerable multitude from many nations, stand at last before the throne of the Lamb of God. Some commentators refer to them as martyrs from the Great Tribulation. Not Wesley! He rightly chose to translate the Greek word *thlipsen* as "affliction" and not "tribulation." So, following the King James translation, Wesley rendered verse 14 - "These are they who come out of great affliction. . ." His comment reads -

> "*These are they* - Not martyrs; for there are not such a multitude as no man can number. But as all the angels appear here, so do all the souls of the righteous who had lived from the beginning of the world. . . *Out of great affliction* - Of various kinds, wisely and graciously allotted by God to all His children."[110]

This grand multitude is portrayed by St. John as ultimately standing before the throne of God, in the divine temple, being busied with "serving" day and night, although then there shall be no night! Dressed in white, symbolic - as Wesley reminds us - of being cleansed from guilt and adorned with holiness, these saints are covered with "the tent" of God's person and protection, never hungering again, nor thirsting any more. The sun's light and heat shall never again afflict them. The Lamb will feed them and will lead them to "living fountains of water," and "God will wipe away all tears from their eyes."[111] Wesley saw this prophetic event in the distant future, coupled with the coming of the New Jerusalem (Revelation XXI and XXII) into the New Creation. Nevertheless, such an event was uttered before hand to provide a teleological context for the breaking of the seventh seal and the beginning of the seven trumpet soundings and the three woes.

The eighth chapter of the Apocalypse opens with - "And when he had opened the seventh seal, there was silence in heaven about half an hour." Wesley observed that silence was "uncommon" in heaven due to the great amount of praise being sounded there, day and night.[112] There is clearly a theological aspect to this silence -

"Immediately before the silence, all the angels, and before them the innumerable had been crying with a loud voice; and now all is still at once: there is a universal pause. Hereby the seal is very remarkably distinguished from the six preceding. This silence before God shows that those who were round about Him were expecting, with the deepest reverence, the great things which the Divine Majesty would further open and order. Immediately after, the seven trumpets are heard, and a sound more august than ever. Silence is only a preparation: the grand point is, the sounding the trumpets to the praise of God."[113]

Moreover, the seven trumpets, to be sounded by seven different angels, announce the progressive activity of God in consummating all things according to his eternal plan. Their sounding is in seven distinct stages, ranging from the time of St. John to the end of the world.[114] The "place of the first four is specified; namely, east, west, south, and north successively: in the last three, immediately after the time of each, the place likewise is pointed out."[115] These four trumpets were sounded in order, without an interval between them. (1) The first was blown and "hail and fire mingled with blood" were cast upon the earth, burning up one third of the earth and one third of the trees with all the grass. Wesley identified the "earth" as here meaning "Asia" and "Palestine in particular."[116] His exegesis of the first trumpet returns the reader's attention back to history -

"Quickly after the Revelation was given, the Jewish calamities under Adrian (Emperor Hadrian, A.D. 117-138) began; yea, before the reign of Trajan was ended. And here the trumpets begin. Even under Trajan, in the year 114, the Jews made an insurrection with a most dreadful fury; And in the parts about Cyrene, in Egypt, and in Cyprus destroyed four hundred and sixty thousand persons. But they were repressed by the victorious power of Trajan, and afterward slaughtered themselves in vast multitudes. The alarm spread itself also into Mesopotamia, where Lucius Quintius slew a great number of them. They rose in Judaea again in the second year of Adrian;

but were presently quelled. Yet in 133 they broke out more violently than ever, under their false messiah Barcochab; and the war continued till the year 135 when almost all Judaea was desolated. In the Egyptian plague also *hail* and *fire* were together. But here *hail* is to be taken figuratively, as also *blood*, for a vehement, sudden, powerful, hurtful invasion; and *fire* betokens the revenge of an enraged enemy, with the desolation therefrom. . . The storm fell, the blood flowed, and the flames raged round Cyrene, and in Egypt and Cyprus, before they reached Mesopotamia and Judaea. . . Fifty well-fortified cities, and eighty-five well-inhabited towns of the Jews were wholly destroyed in this war. Vast tracts of land were likewise left desolate and without inhabitants. . . Some understand by *the trees*, men of eminence among the Jews; by *the grass*, the common people. The Romans spared many of the former: the latter were almost all destroyed. Thus vengeance began at the Jewish enemies of Christ's kingdom; though even then the Romans did not quite escape. But afterwards it came upon them more and more violently: the second trumpet affects the Roman heathens in particular; the third, the dead, unholy Christians; and the fourth, the empire itself."[117]

(2) The second angel sounded his trumpet and "a great mountain burning with fire was cast into the sea; and the third part of the sea became blood; And the third part of the creatures that were in the sea which had life (*psychas* or *souls*), died; And the third part of the ships were destroyed: (Revelation VIII, 8-9). As noted earlier, Wesley linked the second trumpet with the West, or Europe. In his commentary on verses 8 and 9, he interpreted *the sea* to mean "Europe" and "chiefly the middle parts of it, the vast Roman Empire." The reference to *a mountain*, he also claimed, "here seems to signify a great force and multitude of people (Jeremiah LI, 25); so this may point to the irruption of the barbarous nations."[118] To illustrate his point, Wesley cited the warlike Goths who broke in upon Europe about A.D. 250, starting "the irruption of one nation after another," until the form of the Roman Empire was lost while only the name remained.[119] Moreover, Wesley argued, the term fire may mean "the fire of war" and the "rage of those savage nations." As for the "third part" consequences, Wesley held that: (a) the sea becoming blood does not necessarily mean that one third of Europe's population was slain; (b) the creatures of the sea that died did so at the hands of "those merciless invaders;" and (c) the

ships destroyed at sea represent "states or republics" - "And. . . many states were utterly destroyed by those inhuman conquerors! Much likewise of this was literally fulfilled. How often was the sea tinged with blood! How many of those who dwelt mostly upon it were killed! And what number of ships destroyed!"[120]

(3) "And the third angel sounded, and there fell from heaven a great star burning as a torch, and it fell on the third part of the rivers, and on the fountains of waters" (Revelation VIII, 10). In Wesley's geographical scheme, the third trumpet's sounding affects the region south of Patmos - Africa, and Egypt in particular. The prophetic term *rivers* gets exegeted by Wesley as the "Nile" which "overflows every year far and wide." But, it was not the physical phenomenon of flooding that Wesley saw in the passage. Rather, it was the flooding of Christian Egypt with Arianism that concerned him -

> "In the whole African history, between the irruption of the barbarous nations into the Roman Empire, and the ruin of the western empire, after the death of Valentinian the Third, there is nothing more momentous than the Arian calamity, which sprang up in the year 315. It is not possible to tell how many persons, particularly at Alexandria, in all Egypt, and in the neighbouring countries, were destroyed by the rage of the Arians. Yet Afric fared better than other parts of the empire, with regard to barbarous nations, till the governor of it, whose wife was a zealous Arian, and aunt to Genseric, king of the Vandals, was, under that pretense, unjustly accused before the empress Placidia. He was then prevailed upon to invite the Vandals into Afric; who under Genseric in the year 428, founded there a kingdom of their own, which continued till the year 533. Under these Vandal kings the true believers endured all manner of afflictions and persecutions. And thus Arianism was the inlet to all heresies and calamities, and at length to Mahometanism itself."[121]

The reference to a *great star* falling from heaven, Wesley affirmed, applied to "a teacher of the Church" and not "an angel." The star is "Arius" (A.D. 256-336) -

> "One of the stars in the right hand of Christ. Such was Arius. He fell from on high, as it were from heaven, into the most pernicious doctrines, and made in his fall a gazing on all sides, being great, and now burning as a torch. He

fell on the third part of the rivers - His doctrine spread far and wide, particularly in Egypt.

And the name of the star is Wormwood - The unparalleled bitterness, both of Arius himself and of his followers shows the exact propriety of his title. *And the third part of the waters became Wormwood* - A very considerable part of Afric was infected with the same bitter doctrine and spirit. *And many men* (though not a third part of them) *died* - By the cruelty of the Arians."[122]

(4) "And the fourth angel sounded, and the third part of the sun was smitten, and the third part of the moon, and the third part of the stars; so that the third part of them was darkened, and the day shone not for the third part thereof, and the night likewise" (Revelation VIII, 12). Wesley's commentary on this fourth trumpet features three treatments - one of *the sun*, one of *the moon*, and one of *the stars*. *The sun was smitten* -

"Or struck. After the emperor Theodosius died, and the empire was divided into the eastern and the western, the barbarous nations poured in as a flood. The Goths and Huns in the years 403 and 405 fell upon Italy itself with an impetuous force; and the former, in the year 410, took Rome by storm, and plundered it without mercy. In the year 452 Attila treated the upper part of Italy in the same manner. In 455 Valentinian III was killed, and Genseric invited from Afric. He plundered Rome for fourteen days together, Recimer plundered it again in 472. During all these commotions, one province was lost after another, till, in the year 476, Odoacer seized upon Rome, deposed the emperor, and put an end to the empire itself. An eclipse of the sun or moon is termed by the Hebrews, a stroke. Now, such a darkness does not come all at once, but by degrees, so likewise did the darkness which fell on the Roman, particulary the western empire; for the stroke began long before Odoacer: namely, when the barbarians first conquered the capital city."[123]

In addition to the darkening of the sun, "a third part of the moon, and the third part of the stars" were darkened at the sounding of this trumpet. Wesley's commentary emphasizes the similarity of meaning between the "earth, sea, and rivers" of the first through third trumpets and the "sun, moon, and stars" of this fourth trumpet - Both sets of threes represent "the men" that either "inhabit them" or "live under them."[124] These men of the fourth trumpet "are so overwhelmed with calamities in those days of darkness, that they can no longer enjoy the light of heaven: unless it may be thought to imply their being killed; so that the sun, moon, and stars shine to them no more."[125] Such an idea, Wesley held, was borrowed from Ezekiel XXXII, 8 - "I will darken all the lights of heaven over them."[126] Moreover, he argued, as the fourth seal transcended the three preceding seals, so the fourth trumpet transcended the three preceding trumpets. This is so because all who were under the sun, moon, and stars were affected. Nevertheless, the first four trumpet soundings "take up a little less than four hundred years."[127]

The epiphany of an angel between the fourth and fifth trumpet vision - crying, "Woe, Woe, Woe" - was significant to Wesley in that he saw the introduction of three calamities as historically relevant to the Jews. "The three woes. . . stretch themselves over the earth from Persia eastward, beyond Italy westward; all which space had been filled with the gospel by the apostles."[128] While the misery of these three woes were to affect all mankind, "these fell more especially on the Jews."[129] But, Wesley observed, these woes - actually trumpets five, six, and seven - were each introduced with "preludes" of historical events. The first occurred in Persia, in A.D. 454, when Isdegard II "resolved to abolish the Sabbath, till he was, by Rabbi Mar, diverted from his purpose." Twenty years later, under the Persian ruler Phiruz, after intense suffering, many Jews apostatized.[130] The second prelude was the "rise of the Saracens, who, in 510, fell into Arabia and Palestine."[131] The third prelude featured the rise of Pope Innocent I (A.D. 401-417) - and his successors - who "enlarged their episcopal jurisdiction beyond bounds. . . also their worldly power, by taking every opportunity of encroaching upon the empire, which as yet stood in the way of their unlimited monarchy."[132]

(5) "And the fifth angel sounded, and I saw a star" - This star was different from that of Revelation VIII, 11, Wesley claimed. "This star belongs to the invisible world." Wesley's "invisible world" included all types of disembodied spirits, even those of the righteous

dead, but the spirit mentioned in this verse was a "holy angel."[133] This star, "coming swiftly and with great force," fell onto the earth. Upon arrival, he was given the "key of the bottomless pit" - a deep and hideous prison, but different from "the lake of fire."[134] Opening the pit, "there ascended a smoke out of the pit, as the smoke of a great furnace; and the sun and the air were darkened by the smoke. . . and out of the smoke there came forth locusts upon the earth; and power was given them, as the scorpions of the earth have power." Wesley's commentary reads -

"There went forth locusts - A known emblem of a numerous, hostile, hurtful people. Such were the Persians, from whom the Jews, in the sixth century, suffered beyond expression. In the year 540 their academies were stopped, nor were they permitted to have a president for near fifty years. In 589 this affliction ended; but it began long before 540. The prelude of it was about the year 455 and 474: the main storm came on in the reign of Cabades, and lasted from 483 to 532. Toward the beginning of the sixth century, Mar Rab Isaac, president of the academy, was put to death. Hereon followed an insurrection of the Jews, which lasted seven years before they were conquered by the Persians. Some of them were then put to death, but not many; the rest were closely imprisoned. And from that time the nation of the Jews were hated and persecuted by the Persians, till they had wellnigh rooted them out. . . many who were called Christians suffered with them. . . variously tormented."[135]

The description of the locusts (Revelation IX, 7ff), Wesley affirmed, fitted the Persians very nicely -

"This description suits a people neither thoroughly civilized, nor entirely savage; and such were the Persians of that age. . . *the locusts are like horses - With their riders.* The Persians excelled in horsemanship. *And on their heads are as it were crowns - Turbans. And their faces are as the faces of men -* Friendly and agreeable. *And they had hair as the hair of women -* All the Persians of old gloried in long hair. *And their teeth were as the teeth of lions -* Breaking and tearing all things in pieces. *And the noise of their wings . . . -* With their war-chariots, drawn by many horses, they, as it were, flew to and fro. *And they have tails like scorpions -* That is, each tail is like a scorpion, not like the tail of a scorpion. *To hurt men five months -* Five prophetic months; that is, seventy-nine common years. So long did those calamities last. *And they have over them a king* - One by whom they are peculiarly directed and governed. *His name is Abaddon* - Both this and *Apollyon* signify a destroyer. By this he is distinguished from the dragon, whose proper name is Satan."[136]

So, "One woe is past; behold, there come yet two woes after these things." The second woe, the coming of the Saracens, will mean the breaking of Persian power, "under which was the first woe." Wesley saw A.D. 589 as the end of Persian domination in the East. He also noted that Mahomet was twenty-four years of age at that time. Due to spiritual degeneration in Christianity and political chaos in Persia, Mahomet gained the occasion for spreading his new religion and empire.[137] In that same year, Pope Boniface III was declared the universal bishop, and the Church of Rome was affirmed "the head of all churches." Wesley concluded - "This was a sure step to advance the Papacy to its utmost height."[138] The first woe, or fifth trumpet, was completed then by 589. The second and third woes, or sixth and seventh trumpets, "quickly followed."[139]

(6) The sixth trumpet sounded and the prophet John reported - "I heard a voice from the four corners of the golden altar which is before God, saying to the sixth angel who had the trumpet, Loose the four angels who are bound in the great river Euphrates" (Revelation IX, 13ff). In his exegesis, Wesley identified the "golden altar" as the "heavenly (Platonic?) pattern of the Levitical altar of incense" in cultic Judaism.[140] As for the voice, Wesley believed it was an expression "of the wrath of God." Divine wrath should be withheld no longer. Since the prayers of saints arise to heaven as "incense" and arrive at the golden altar, it is appropriate that the divine voice should speak retribution upon evil persecutors of believers from the four corners of that altar. ". . .the execution of the wrath of God. . . should, at no intercession, be delayed any longer."[141] As for the "four angels" ordered loosed, Wesley argued they were "evil angels," or they would not have been bound (imprisoned). "Why, or how long, they were bound we know not."[142] Nonetheless, they were loosed, and each going forth with

strength and rage, they set about "to kill the third part of men." Wesley qualified this last phrase - "That is, an immense number of them"[143] The time dimension for this slaughter, according to John the Presbyter, involved the "hour, and day, and month, and year." This time sequence must be rightly understood, Wesley argued, if one is to gain historical appreciation for what God has already done in preparing for the consummation of all things in Christ. To simplify his argument, Wesley provided a chart of prophetic time -

	Common Years	Common Days	
Hour		8	
Day		196	In all,
Month	15	318	212 years
Year	196	117	

How does Wesley use these 212 common years, drawn from prophetic time? First, he interpreted the four angels as the caliphs who succeeded Mahomet, then he treated the 212 years -

"All this agrees with the slaughter which the Saracens made for a long time after Mahomet's death. And with the number of angels let loose agrees with the number of their first and most eminent caliphs. These were Ali, Abubeker, Omar, and Osman. Mahomet named Ali, his cousin and son-in-law, for his successor; but he was soon worked out by the rest, till they severally died, and so made room for him. They succeeded each other, and each destroyed innumerable multitudes of men. . . from their first caliph, Abubeker, till they were repulsed from Rome under Leo IV. These 212 years may therefore be reckoned from the year 632 to 847. The gradation in reckoning the time, beginning with the hour and ending with the year, corresponds with their small beginning and vast increase. Before and after Mahomet's death, they had enough to do to settle their affairs at home. Afterwards Abubeker went farther, and in the year 634 gained great advantage over the Persians and Romans in Syria. Under Omar was the conquest of Mesopotamia, Palestine, and Egypt made. Under Osman, that of Afric (with the total suppression of the Roman government in the year 647), of Cyprus, and of all Persia in 651. After Ali was dead, his son, Ali Hasen, a peaceable prince, was driven out by Muavia; under whom, and his successors, the power of the Saracens so increased, that within fourscore years after Mahomet's death they had extended their conquests farther than the warlike Romans did in four hundred years."[144]

Wesley understood the horsemen of the sixth trumpet vision to be symbolic of early Islam. Revelation's "two hundred millions" of riders equal the Islamic forces that swept the world during the 212 years cited. So, the "third part of men" were not killed all at once, but "during the course of years."[145] The caliph Omar "alone, in eleven years and a half, took thirty-six thousand cities or forts. How many men must be killed therein!" Wesley exclaimed.[146] The countries taken by Islam, Wesley observed, were those where the gospel had been planted. Though "plagued" by such conquests, the Christians did not repent of the growing evil of venerating the images of departed saints, which the second Council of Nicea authorized in A.D. 787 - a clear instance, for Wesley, that councils of men can and do err.[147] In Wesley's words - "Here the description of the second woe ends."

The seventh trumpet, with the third woe, is not presented in the Apocalypse until after an interlude which consists of chapter X, 1 through chapter XI, 13. However, the important consideration at the close of Wesley's treatment of the sixth trumpet (Revelation IX) is his emphasis that chapters VI through IX involve prophecies about signs of the end that have already been fulfilled in history. There are three more divisions of the Apocalypse left for Wesley to explain - Chapters X - XIV, things now being fulfilled; Chapters XV - XIX, things to be fulfilled shortly; and Chapters XX - XXII, things at a greater distance.

D. Things now being fulfilled - From verse one of chapter X through chapter XI, 13, the apocalyptic author was setting the stage for the dramatic sounding of the seventh trumpet and the arrival of the third great woe. Wesley realized this, and he identified the passage as having two parallel parts: (1) Chapter X, 1 - 7; and (2) Chapters X, 8 - XI, 13. The time period covered by these overlapping events, as Wesley calculated, was lengthy, starting during the second woe (sixth trumpet) and reaching into the activities of the third woe (seventh trumpet). However, many things revealed in them "are not fulfilled till long after."[148]   The "consummation of the mystery of God" (X, 7) is an example of things "not fulfilled till long after." Also, the ascent of the beast "out of the bottomless pit" (XI, 7) is an event yet to come.[149] Wesley's exegesis of the parallel passages unfolds as follows: (1) Revelation X, 1 - 7. Verse 1 - "Another mighty angel" - Wesley distinguished this angel from the other mighty angel of Revelation V, 2. The rainbow upon his head shows divine favor, and the cloud that clothed him indicates "his high dignity."[150] His face, as bright as the sun, witnesses to his great righteousness. The angelic feet, as pillars of fire, bespeak the purifying presence of Deity, swift to purge every *topos*. Wesley's commentary is interesting at this point -

> "*And he had in his hand* - His left hand: he swore with his right. He stood with his right foot on the sea, toward the west; his left, on the land, toward the east; so that he looked southward. And so St. John (as Patmos lies near Asia) could conveniently take the book out of his left hand. This sealed book was first in the right hand of Him that sat on the throne: thence the Lamb took it, and opened the seals. And now this little book, containing the remainder of the other, is given opened, as it was, to St. John. From this place the Revelation speaks more clearly and less figuratively than before. *And he set his right foot upon the sea* - Out of which the first beast was to come. *And his left foot upon the earth* - Out of which was to come the second. *The sea* may betoken Europe: *the earth*, Asia; the chief theatres of these great things."[151]

As this mighty angel cried with a loud voice, "as a lion roareth," seven "thunders uttered their voices' (verse 3). Wesley claimed these "thunders" were "glorious, heavenly powers, whose voice was as the loudest thunder." These voices - of the mighty angel and of the seven powers - while very loud, were but a prelude to a louder and more commanding voice -

> "*And I heard a voice from heaven* - Doubtless from Him who had at first commanded him to write, and who presently commands him to take the book; namely, Jesus Christ. *Seal up those things which the seven thunders have uttered, and write them not* - These are the only things of all which he heard that he is commanded to keep secret: so something peculiarly secret

was revealed to the beloved John, besides all the secrets that are written in this book. At the same time we are prevented from inquiring what it was which these thunders uttered; suffice that we may know all the contents of the opened book, and of the oath of the angel."[152]

Verses 5 and 6 (Revelation X) depict the mighty angel in the act of swearing an oath. The oath is simple - "There shall be no more a time." This oath was one of the most important bits of biblical eschatology for John Wesley. Indeed, his understanding of the consummation of all things in Christ depends upon the correct interpretation of this oath. Rightly deciphered, Wesley believed, an approximate date for "the end" is obvious - A.D. 1836! His commentary on this oath is quite explicit -

"*And sware* - The six preceding trumpets pass without any such solemnity. It is the trumpet of the seventh angel alone which is confirmed by so high an oath. *By Him that liveth for ever and ever* - Before whom a thousand years are but a day. *Who created the heaven, the earth, the sea, and the things that are therein* - And, consequently, has the sovereign power over all; therefore, all His enemies, though they rage a while in heaven, on the sea, and on the earth, yet must give place to Him. *That there shall be no more a time* - But in the days of the voice of the seventh angel, the mystery of God is fulfilled: that is, *a time, a chronos,* shall not expire before that mystery is fulfilled. a *chronos* (1,111 years) will nearly pass before then, but not quite. The period, then, which we may term a *non-chronos* (not a whole time), must be a little, and not much, shorter than this. The *non-chronos* here mentioned seems to begin in the year 800 (when Charles the Great instituted in the West a new line of emperors or of many kings); to end in the year 1836; and to contain, among other things, the 'short time' of the third woe, the 'three times and a half' of the woman in the wilderness, and the 'duration' of the beast."[153]

Wesley's treatment of verse 7 emphasizes the special significance of the phrase "the mystery of God." The text claims that the sounding of the seventh trumpet will mean the fulfilling of "the mystery of God' as declared to the prophets. John of Epworth observed - "It is said, Rev. XVII, 17, 'The word of God shall be fulfilled.' The word of God is fulfilled by the destruction of the beast; the *mystery*, by the removal of the dragon."[154] As yet, Wesley has not treated the subjects of the "beast" and the "dragon." But, their time is near! Until then, it must be noted that Wesley saw the "destruction of the beast" and the "removal of the dragon" as almost being one event, although technically they are two. They occupy the same segment of time - they have their beginning in heaven, as soon as the seventh trumpet sounds, and their end on earth and sea. The angel's oath declares that they shall be terminated. Such a declaration, Wesley asserted, is a "comfort" for "holy men" - Christians "who are afflicted under that woe."[155] It is clear from this tenet that Wesley would reject the modern doctrine of the "Rapture," whose advocates generally argue that "God would never let his saints suffer during the Great Tribulation!" Wesley's understanding of the Apocalypse has the Church suffering continuously from the rise of Domitian to 1836. Hence, the oath of the angel serves as a great comfort to "holy men" of all ages, but especially to those living during the time of the seventh trumpet sounding and the third woe. However, to the unholy there is nothing comforting about the oath and the mystery of God - "Indeed, the wrath of God must first be fulfilled, by the pouring out of the phials; and then comes the joyful fulfilling of the mystery of God."[156] Wesley added this point - "To the seventh trumpet belongs all that occurs from Rev. XI, 15 to XXII, 5." The third woe, as part of the seventh trumpet, is described in Revelation XII, 12 and XIII, 18.[157]

(2) Chapter X, 8 - XI, 13. According to Wesley, this passage runs parallel to chapter X, 1 -7. However, the symbolism is considerably different. For instance, in chapter X, 9, the prophet John is commanded to eat the book previously delivered to him by the mighty angel. He obeys, and the book is sweet in his mouth, then bitter in his belly, like Ezekiel's experience centuries earlier. Wesley observed - "This was an emblem of thoroughly considering and digesting it. . . The sweetness betokens the many good things which follow (Rev. XI, 1, 15, etc.); the bitterness, the evils which succeed under the third woe."[158] The prophet is commanded to "prophesy again" (verse 11) concerning "people, and nations, and tongues, and many kings." The people and nations, for Wesley, "are contemporary," but the kings, being many, succeed one an-

other, and they are related to the holy city (Revelation XI, 2),
which is old Jerusalem. John's measuring of the temple in
Jerusalem resembles the project of Ezekiel, except John is forbidden to measure the court of the Gentiles (verse 2) because the
Gentiles will "tread the holy city forty-two months." These Gentile
kings, in Wesley's scheme, have been both Christian and Turkish,
ruling over Jerusalem for centuries.[159] The forty-two months, surprisingly, must be taken as common months rather than figurative
months. Wesley also reduced these months to common days -
twelve hundred and sixty of them.[160] These months or days are
part of the "non-chronos" period of the seventh trumpet sounding.
Another theme struck in this parallel passage is that of Christ's two
witnesses - "two select, eminent instruments. Some have supposed
(though without foundation) that they are Moses and Elijah, whom
they resemble in several respects."[161] These two prophesy twelve
hundred and sixty days, common days, according to Wesley. Their
kerygma announces that Jesus is the Son of God, the heir of all
things, and they exhort everyone to repent, and fear, and glorify
God. Being dressed in sackcloth, they prophesy out "of sorrow and
concern for the people." They are symbolically identified as "olive
trees" - a symbol taken from Zechariah III, 9, meaning "two chosen
instruments," full of the unction of the Holy One.[162] They are also
called "candlesticks" because of their bright and shining witness.
Wesley noted that they are "always waiting on God, without the
help of man, and asserting His right over the earth and all things
therein." Moreover, if anyone should kill them, that one would be
killed by "devouring fire." These two prophets have great power,
and, Wesley argued, they use that power. As Moses turned the
Nile into blood, these prophets have the power to turn the water
near Jerusalem into blood. They, unlike Moses and Elijah, have
power to smite the earth with plagues "as they will." They are
"invincible" until they have finished their testimony.[163] Enter now
the "beast" (Revelation XI, 7) - the "wild beast that ascendest out
of the bottomless pit." Being loosed onto the earth, this vile monster "makes war on the two witnesses." Wesley affirmed that "that
time ends after the ascent of the beast out of the abyss, and yet be-

fore the fulfilling of the mystery."[164] The two prophets are conquered by the beast and killed, making them among the last martyrs
of God, although they are not the last. Their dead bodies lie in the
street of Jerusalem. However, Wesley thought they might be hung
on crosses. Wesley, rich with alternative ideas, suggested that the
exposing of the corpses of the two witnesses might take place
within the Church of the Holy Sepulcher.[165] For three and a half
common days - Wesley said, "Friday, Saturday, and Sunday" - the
people of the heathen nations have "leisure to gaze upon and rejoice over them."[166] "Earthly minded men" make merry over them,
"as did the Philistines over Sampson. And send gifts to one another
- Both Turks, and Jews, and heathens, and false Christians."[167] At
the end of three and a half days, Wesley observed, the "spirit of life
from God came into them," that is, into the dead bodies of the two
witnesses, "and they stood on their feet." The result was pre-
dictable - "Great fear fell upon them that saw them." A loud voice
from heaven sounded and ordered the two prophets to "Come up
hither." As Jesus ascended into heaven in a cloud (Acts I, 9), so
these two witnesses "went up into heaven in a cloud; and their enemies beheld them" (Revelation XI, 12). Immediately, "a tenth
part of the city fell" because of a great earthquake.[168]

A brief digression is necessary here. Wesley was enthralled with
the subject of earthquakes. He viewed them as divine acts. In his
famous sermon, *THE CAUSE AND CURE OF EARTHQUAKES*,
he introduced the subject with these awesome words - "Of all the
judgments which the righteous God inflicts on sinners here, the
most dreadful and destructive is an earthquake."[169] Moreover,
they are set forth by the inspired authors of scripture "as God's
proper judicial act, or the punishment of sin: Sin is the cause,
earthquakes the effect, of his anger."[170] While earthquakes bring
punishment for sin, they frequently have a converting effect upon
the survivors. This digression is pertinent to our survey of Wesley's
exegesis of the Apocalypse after all. In chapter XI, verse 13, the
great earthquake that destroys a tenth part of the city of Jerusalem,
after the ascension of the two witnesses into heaven, takes the lives
of seven thousand men, but the survivors, being terrified, "gave

glory to the God of heaven." Wesley's commentary speaks in terms of conversion -

"We have here an unanswerable proof that this city is not Babylon or Rome, but Jerusalem. For Babylon shall be wholly burned before the fulfilling of the mystery of God. But this city is not burned at all; on the contrary, at the fulfilling of that mystery, a tenth part of it is destroyed by an earthquake, and the other nine parts converted. *And there were slain in the earthquake seven thousand men* - Being a tenth part of the inhabitants, who therefore were seventy thousand in all. *And the rest* - The remaining sixty-three thousand were converted: a grand step toward the fulfilling of the mystery of God. Such a conversion we nowhere else read of. So there shall be a larger as well as holier church at Jerusalem than ever was yet. *Were terrified* - Blessed terror! *And gave glory* - The character of true conversion (Jer. XIII, 16). *To the God of heaven* - He is styled 'The Lord of the earth' (verse 4) when He declared His right over the earth by the two witnesses; but *the God of heaven*, when He not only gives rain from heaven after the most afflicting drought, but also declares His majesty from heaven, by taking His witnesses into it. When the whole multitude gives glory to the God of heaven, then that 'treading of the holy city' ceases."[171]

With the above words the parallel passages - chapter X, 1 - 7 and chapters X, 8 through XI, 13 - come to an end in Wesley's *EX-PLANATORY NOTES*. "This is the point (in history) so long aimed at, the desired 'fulfilling of the mystery of God,' when the divine promises are so richly fulfilled on those who have gone through so great afflictions."[172] Yet, Wesley was far from the end of his commentary on this division of Revelation (chapters X through XIV). After chapter XI, verse 13, there are still many things being fulfilled. A running account of chapter XI, 14ff through chapter XI, with Wesley's notes, is profitable.

Chapter XI, 14 - "The second woe is past; behold, the third woe cometh quickly." The second woe came to an end about the year 847, when Charles the Great broke the military power of the Saracens, bringing a halt to their butchery. The third woe began with the Vatican stealing universal authority, "enlarging its power and grandeur." In the year 755, the bishop of Rome (Stephen II) became a secular prince when King Pepin gave him the exarchate of Lombardy.[173]

Chapter XI, 15 - "And the seventh angel sounded" - At long last, the seventh trumpet is heard. This sounding, Wesley held, heralds the "most important and joyful events," surpassing everything preceding it. All seven trumpet soundings were heard in heaven, but the seventh is heard on earth as well. Wesley linked this "last trumpet" with I Thessalonians IV, 16 - "The Lord himself will descend from heaven with a shout, with the voice of an archangel, and with the trumpet of God." Moreover, the reference to the resurrection at Christ's coming (I Corinthians XV, 52) speaks emphatically of this trumpet being heard on earth.[174] In the Apocalypse, as soon as the seventh trumpet is heard, "the kingdom falls to God and His Christ." This is first witnessed in heaven, "and is there celebrated with joyful praise." However, "on earth several dreadful occurrences are to appear first." The events on earth, following this last sounding trumpet, are related from this verse through chapter XXII, 5. Reference to "The kingdom of the world" is made here, emphasizing "royal government over the whole world," under the power of Satan (the dragon). But with the seventh trumpet, this kingdom reverts back to its rightful master, God. The divine agent to consummate this reversion is Jesus, the Christ. The term *christ* means "anointed" and refers to a king. Hence, at the sound of the last trumpet, "the kingdom of this world is become the kingdom of our Lord and of his Christ." Wesley added, "Yet Satan and the present world, with its kings and lords, are risen against the Lord and His Anointed. God now puts an end to this monstrous rebellion, and maintains His right to all things. And this appears in an entirely new manner, as soon as the seventh angel sounds."[175]

Chapter XI, 16 - 19 - With the sounding of the seventh trumpet, the twenty-four elders - introduced in chapter IV, 4 - fall down on their faces in heaven and begin to worship God, saying -

"We give thee thanks, O Lord God, the Almighty, who is, and who was; because thou hast taken the great power, and hast reigned. And the nations were wroth, and thy wrath is come, and the time of the dead, that they be judged, and to give a reward to thy servants the prophets, and to the saints, and to them that fear thy name, small and great; and to destroy them that destroyed the earth."[176]

While Wesley gave an interpretive meaning to the various phrases of the elders' chant, the most significant one treated involves the last line - "to destroy them that destroy the earth." The contemporary reader might easily slip a line or two into Wesley's, showing greed spoiling the earth's environment -

"The earth was destroyed by the great whore in particular (Rev. XIX, 2; XVII, 2, 5); but likewise in general, by the open rage and hate of wicked men against all that is good; by wars, and the various destruction and desolation naturally flowing therefrom; by such laws and constitutions as hinder much good, and occasion many offences and calamities; by public scandals, whereby a door is opened for all dissoluteness and unrighteousness; by abuse of secular and spiritual powers; by evil doctrines, maxims, and councils; by open violence and persecution; and by sins crying to God to send plagues upon the earth. This great work of God, destroying the destroyers, under the trumpet of the seventh angel, is not the third woe, but matter of joy, for which the elders solemnly give thanks. All the woes, and particularly the third, go forth over those who dwell upon the earth; but this destruction is over those 'who destroy the earth,' and were also instruments of that woe."[177]

The prophet John concludes this passage with reference to the heavenly temple - "And the temple of God was opened in heaven, and the ark of the covenant was seen in the temple; and there were lightnings, and voices, and thunders, and an earthquake, and great hail." Wesley observed that the lightnings, voices, thunders, earthquake, and hail, originating in the heavenly temple, are later poured out of phials onto the earth by the seventh angel (Revelation XVI, 17 - 21). "What the trumpet here denounces in heaven is there executed by the phials upon earth. First it is shown what will be done; and afterwards it is done."[178]

While we have not completed our study of Wesley's treatment of chapters X through XIV - things now being fulfilled - the major signs of the end and Christ's coming have been explored. Those signs that remain untreated, at this juncture of the study, will be encountered in the next chapter of this work.

The Methodist pilgrim, living within the tension of the seventh trumpet events, must recognize the signs already fulfilled. Pilgrim patience, loving stewardship, and steadfast endurance must increase. Suffering and tears must not dim the bright expectation of Christ's coming. Pilgrim times are in His hands, and by His hand the great *telos* is attained. Pilgrims are especially blessed because they mourn, and in the consummation they shall be comforted. Wesley translated a German Pietistic hymn by Paul Gerhardt that addresses this great truth -

> "Give to the winds thy fears;
> Hope, and be undismayed;
> God hears thy sighs and counts thy tears;
> God shall lift up thy head.
>
> Through waves, and clouds, and storms,
> He gently clears thy way;
> Wait thou His time, so shall this night
> Soon end in joyous day.
>
> Still heavy is thy heart?
> Still sink thy spirits down?
> Cast off the weight, let fear depart,
> And every care be gone.
>
> What though thou rulest not?
> Yet heaven, and earth, and hell
> Proclaim, 'God sitteth on the throne,
> And ruleth all things well!'
>
> Leave to His sovereign sway
> To choose and to command:
> So shalt thou, wondering, own His way,
> How wise, how strong His hand!
>
> Far, far above thy thought
> His counsel shall appear,
> When fully He the work hath wrought
> That caused they needless fear![179]

Notes

1   Telford, *op.cit.*, VII, 358, VIII, 251.
2   EXPLANATORY NOTES NEW TESTAMENT, *op.cit.*, 792.
3   *Ibid*, 908.
4   *Ibid*, 985.
5   *Ibid*, 112.
6   *Ibid*, 112-113.
7   *Ibid*, 113.
8   *Loc.cit.* - Wesley's reference to being "snatched out" should not be confused with the popular doctrine of the Rapture. Wesley did not know that doctrine, nor did Calvin, Luther, Aquinas, Augustine, Cyprian, Tertullian, or Paul. It was the product of a small English sect in the mid-nineteenth century.
9   *Loc.cit.*
10   *Loc.cit.*
11   *Ibid*, 114.
12   *Loc.cit.*
13   *Loc.cit.*
14   *Loc.cit.*
15   *Ibid*, 115.
16   *Loc.cit.*
17   *Loc.cit.*
18   *Ibid*, 898.
19   *Ibid*, 115.
20   *Loc.cit.*
21   *Ibid*, 116.
22   *Loc.cit.*
23   *Loc.cit.*
24   *Ibid*, 117.
25   *Loc.cit.* Cf, WORKS, *op.cit.*, VII, 111. In his sermon *ON OBEDIENCE TO PASTORS*, Wesely declared - "I dare not receive one as my guide to heaven, that is himself in the high road to hell. I dare not take a wolf for my shepherd...that is a common swearer, an open drunkard, a notorious sabbath-breaker."
26   *Ibid*, 760.
27   *Ibid*, 766.
28   *Ibid*, 767.
29   *Loc.cit.*
30   *Loc.cit.*
31   WORKS, *op.cit.*, VI, 254.

32   *Ibid*, VI, 256.
33   *Loc.cit.*
34   *Ibid*, VI, 257.
35   *Ibid*, VI, 258.
36   *Loc.cit.*
37   *Ibid*, VI, 259.
38   *Loc.cit.*
39   *Loc.cit.*
40   *Ibid*, VI, 261.
41   *Loc.cit.*
42   *Ibid*, VI, 262.
43   *Loc.cit.*
44   *Ibid*, VI, 263. Cf. *Ibid*, VI, 329. The dying words of Luther confirm this short-coming - "I have spent my whole life for nought! Those who are called by my name, are, it is true, reformed in opinions and modes of worship; but in their hearts and lives, in their tempers and practice, they are not a jot better than the Papists."
45   *Loc.cit.*
46   *Ibid*, VI, 264.
47   *Loc.cit.*
48   EXPLANATORY NOTES NEW TESTAMENT, *op.cit.*, 908.
49   *Ibid*, 932.
50   *Loc.cit.*
51   *Loc.cit.*
52   *Ibid*, 933.
53   *Ibid*, 934.
54   *Loc.cit.*
55   *Loc.cit.*
56   *Ibid*, 935.
57   *Loc.cit.*
58   *Ibid*, 954.
59   *Ibid*, 936.
60   *Loc.cit.*
61   *Loc.cit.*
62   *Ibid*, 937.
63   *Loc.cit.*
64   *Ibid*, 938.
65   *Loc.cit.*
66   *Ibid*, 939.
67   *Ibid*, 941.
68   *Loc.cit.*
69   *Ibid*, 942.
70   *Ibid*, 944.
71   *Ibid*, 942.
72   *Ibid*, 944.
73   *Ibid*, 946.

74   *Ibid*, 947.
75   *Ibid*, 948-949.
76   *Ibid*, 950.
77   *Ibid*, 951.
78   *Ibid*, 952-953.
79   *Ibid*, 944-945.
80   *Ibid*, 954.
81   *Ibid*, 955. It is significant that the renovation of Wesley Chapel in London featured replacement of the ship-mast columns (given by King George III) with jasper ones (given by the American Methodists).
82   *Loc.cit.*
83   *Ibid*, 956.
84   *Ibid*, 958.
85   *Ibid*, 959.
86   *Loc.cit.*
87   *Loc.cit.*
88   *Ibid*, 961.
89   *Ibid*, 963.
90   *Loc.cit.*
91   *Loc.cit.*
92   *Ibid*, 964.
93   *Ibid*, 964-965.
94   *Ibid*, 965.
95   *Ibid*, 966.
96   *Loc.cit.*
97   *Loc.cit.*
98   *Ibid*, 967.
99   *Loc.cit.*
100   *Loc.cit.*
101   *Ibid*, 968.
102   *Ibid*, 968-969.
103   *Ibid*, 968.
104   *Ibid*, 969.
105   *Loc.cit.*
106   *Loc.cit.*
107   *Ibid*, 970.
108   *Loc.cit.*
109   *Loc.cit.*
110   *Ibid*, 972.
111   *Loc.cit.*
112   *Loc.cit.*
113   *Ibid*, 973. Wesley suggested that the "half an hour" be taken literally.
114   *Loc.cit.*
115   *Loc.cit.*
116   *Ibid*, 975.
117   *Loc.cit.*

118   *Ibid*, 975-976.
119   *Ibid*, 976.
120   *Loc.cit.*
121   *Loc.cit.*
122   *Ibid*, 977.
123   *Loc.cit.*
124   *Loc.cit.*
125   *Loc.cit.*
126   *Loc.cit.*
127   *Ibid*, 978.
128   *Loc.cit.*
129   *Loc.cit.*
130   *Loc.cit.*
131   *Loc.cit.*
132   *Loc.cit.*
133   *Loc.cit.*
134   *Ibid*, 979.
135   *Loc.cit.*
136   *Ibid*, 980.
137   *Ibid*, 980-981.
138   *Ibid*, 981.
139   *Loc.cit.*
140   *Loc.cit.*
141   *Loc.cit.*
142   *Loc.cit.*
143   *Loc.cit.*
144   *Ibid*, 981-982.
145   *Ibid*, 982.
146   *Loc.cit.*
147   *Loc.cit.*
148   *Ibid*, 983.
149   *Ibid*, 984.
150   *Loc.cit.*
151   *Loc.cit.*
152   *Ibid*, 984-985.
153   *Ibid*, 985.
154   *Ibid*, 986.
155   *Loc.cit.*
156   *Loc.cit.*
157   *Loc.cit.*
158   *Loc.cit.*
159   *Ibid*, 987.
160   *Loc.cit.*
161   *Ibid*, 988.
162   *Loc.cit.*
163   *Loc.cit.*

164 *Loc.cit.*
165 *Ibid*, 989.
166 *Loc.cit.*
167 *Loc.cit.*
168 *Loc.cit.*
169 WORKS, *op.cit.*, VII, 386.
170 *Ibid*, VII, 387.
171 EXPLANATORY NOTES NEW TESTAMENT, *op.cit.*, 990.
172 *Loc.cit.*
173 *Ibid*, 991.
174 *Ibid*, 639.
175 *Ibid*, 991.
176 *Ibid*, 992-993.
177 *Ibid*, 993.
178 *Loc.cit.*
179 METHODIST HYMNAL, *op.cit.*, 673.

Chapter Ten

The End and the Beginning: The Final Signs

On November 3, 1772, John Wesley asked his Methodist pilgrims a pertinent question - "Is not the Lord at hand?"[1] They knew the correct answer. For many years, Wesley had instructed them on such matters. Colin Williams has captured the spirit of Wesley's eschatological teachings, showing how the kingdom of God had a dimension of "now" and another of "not yet." Williams comments - "Wesley sees the Christian life as a pilgrim way, with the present time as a time of decision and preparation for the final life of the kingdom still to come."[2] The right decision can be made because the kingdom is now, and enabling grace is freely given as a consequence of the kingdom's presence. However, the glories of the kingdom of God, as they now apply, cannot be compared to the glories of the kingdom consummated and yet to come. In his commentary on the Epistle to the Hebrews II, 8-9, Wesley observed -

"Thou hast put all things in subjection under his feet. Now in putting all things in subjection under him, he left nothing that is not put under him. But now we do not see all things put under him. But we see Jesus crowned with glory and honour, for the suffering of death, who was made a little lower than the angels, that by the grace of God he might taste death for every man.

. . .it is plain this is done not now, with regard to man in general. It is done only with regard to Jesus, God-Man, who is now crowned with *glory and honour*."[3]

In so far as pilgrims are concerned, some of the "now" are "not yet." There are still many things not put under their feet - temptations, ignorance, principalities and powers, poor health, pain, and death, to cite a few. But, Wesley believed, in regard to Christ Jesus "all things" are now under his feet by virtue of his sacrificial death and God's reception of him with "glory and honour." While the pilgrim may not sense that Christ is trampling the kingdom of darkness, *HE IS*, in a *Deus Absconditus* sort of way. Nevertheless, Wesley held, there are revelatory indications of the kingdom of God hastening onward to full consummation, if only pilgrims have spiritual eyes to see and believe. Hence, the kingdom of God is not "merely an inward reality, cut off practically from surrounding culture. . . Wesley looked forward to the strange sight of a Christian world, foreseen by the prophets and apostles."[4]

In the previous chapter, Wesley's tracing of divine activities as signs of the future consummation of the kingdom of God were drawn from Gospel, Epistle, and the first eleven chapters of the Apocalypse. The analytical treatment of the book of Revelation now must be resumed, beginning with chapter twelve.

THE FINAL SIGNS OF THE APOCALYPSE

In Wesley's scheme of interpretation, chapters twelve through fourteen treat "things which are now fulfilling." Chapters fifteen through nineteen treat "things which will be fulfilled shortly."[5] Our examination here will consider these two blocks of prophetic material as Wesley understood them.

A. *Things Now Being Fulfilled* (Revelation XII-XIV).
Revelation XII - This remarkable chapter opens with the solemn words - "And a great sign was seen in heaven." Wesley's definition of a "sign" is short - "something that has an uncommon appearance, and from which we infer that some unusual things will follow."[6] While short, it is pregnant with meaning because of the term "infer" - a term of logic! A sign does not reveal its truth with a burst of epignosis into the mind of its observer. Never! A sign is an indication of something else. It points beyond itself to a non-apparent referent. Wesley clearly preempted Paul Tillich at this juncture.

The way from sign to referent is "inference" as an act of reason. For Wesley it was an act of Christian reason. Obvious and literalistic interpretations of signs are not to be taken seriously. Hence Wesley rejected the popular idea that the referent of this sign of the heavenly woman is the Virgin Mary. Instead, he, as others in the Anglican tradition, inferred it to mean the "Church." The text concerning the sign reads -

> "And a great sign was seen in heaven; a woman clothed with the sun, and the moon under her feet, and on her head a crown of twelve stars: And being with child she crieth, travailing in birth, and pained to be delivered." (vss. 1-2)

Wesley's commentary explains the meaning of the "woman" as the Church -

> "*A woman* - The emblem of the Church of Christ, as she is originally of Israel, though built and enlarged on all sides by the addition of heathen converts; and as she will hereafter appear when all her 'natural branches' are again 'grafted in.' She is at present on earth: and yet, with regard to her union with Christ, may be said to be in heaven (Eph. II, 6). Accordingly, she is described as both assaulted and defended in heaven (verses 4, 7). *Clothed with the sun, and the moon under her feet, and on her head a crown of twelve stars* - These figurative expressions must be so interpreted as to preserve a due proportion between them. So, in Joseph's dream, the sun betokens his father; the moon, his mother; the stars, their children. There may be some such resemblance here: and as the prophecy points out the 'power over all nations,' perhaps the *sun* may betoken the Christian world; the *moon*, the Mahometans, who also carry the moon in their ensigns; and the *crown of twelve-stars*, the twelve tribes of Israel; which are smaller than the sun and the moon. The whole of this chapter answers the state of the Church from the ninth century to this time."[7]

Since the sign of the woman in travail represents the Church in Wesley's system, what meaning is to be given for her cries? They are "painful longings, the sighs, and prayers of the saints for the coming kingdom of God." Restated, "the woman groaned and travailed in spirit, that Christ might appear, as the Shepherd and King of all nations."[8]

Another sign immediately appeared in heaven - a great red dragon with seven heads, ten horns, and seven diadems on his heads (vss. 3-4). His tail drew a third of the stars together and cast them to the earth, and he stood before the woman who was ready to deliver her child, intending to "devour the child." She brought forth a man-child, "who was to rule all the nations with a rod of iron: and her child was caught up to God, and to his throne." The woman fled into the wilderness, where God had prepared a place for her to be fed for twelve hundred and sixty days (vss. 5-6).

What meanings did Wesley give to the various elements of this sign? First and foremost, the great red dragon, as the prophet John declares in verse 9, is the "ancient serpent, who is called the Devil, and Satan, who deceiveth the whole world." His fiery-red color, Wesley asserted, typifies his disposition, and his seven heads indicate vast wisdom. The ten horns represent "mighty power and strength, which he still retained." His seven diadems are not "crowns" but "costly bindings, such as kings anciently wore." Though a fallen creature, the dragon "was a great potentate still, even the prince of this world."[9] The tail of the dragon is symbolic of "falsehood and subtilty." By such lies and deceit, Satan drew "a third part" (a very large number) of "stars of heaven" (Christians and their teachers who once sat in heavenly places with Christ) and "cast them to the earth" (utterly depriving them of all those heavenly blessings). This great deception and casting down took place between the years A.D. 847 and 947, when the Manichees in the east "drew abundance of people from the truth." Thus the Devil sought to "hinder the kingdom of Christ from spreading abroad," and this is what is meant by the phrase "he might devour the child."[10]

The second element of this sign is the child - a man-child, even Christ, not in his person but in his kingdom. The Church brought forth a great expansion of Christ's kingdom, "In the ninth age, many nations with their princes were added to the Christian Church."[11] The child, who is to rule over all nations, shall come when his time arrives. Until then, both the woman and her son are in heaven, having been caught up to God, taken completely out of the reach of the great dragon.[12] One should not take Wesley's "caught up" interpretation to support the modern idea of a "rapture." For Wesley the Church that is presently in heaven (Abraham's Bosom) is the Church Triumphant, or the righteous dead of all time. The Church Militant is still on earth, as the woman fleeing in the wilderness (verse 6).

The "wilderness" is the earth, and more specifically Europe - "As Asia and Afric were wholly in the hands of the Turks and Saracens; and in a part of it where the woman had not been before." God's preparation of "a place" for the Church refers to a place made "safe and convenient for her."[13] The total area of Europe, from the Danube River westward, from the ninth century, was given by God to the Church. The people of Europe then began to provide "all things needful for her." The twelve hundred and sixty days (verse 6) are "prophetic days" which Wesley deciphered into "seven hundred and seventy-seven common years." Following Bengelius, Wesley added the 777 to A.D. 847 and arrived at the date of A.D. 1524 - the end of a period of Church conquest on earth, in Europe - "So long the woman enjoyed a safe and convenient place in Europe, which was chiefly Bohemia; where she was fed, till God provided for her more plentifully at the Reformation."[14]

Verse 7 begins a description of war in heaven between Michael and his angels and the great dragon and his angels, but the dragon did not prevail and his place in heaven was not "found any more." He was cast to the earth and his angels with him. And the prophet John heard the proclamation from heaven -

"Now is come the salvation, and the might, and the kingdom of our God, and the power of his Christ: for the accuser of our brethren is cast out, who accused them before our God day and night. And they have overcome him by the blood of the Lamb, and by the word of their testimony; and they loved not their lives unto the death. Therefore rejoice, ye heavens, and ye that dwell in them. Woe to the earth and the sea! for the devil is come down to you, having great wrath, because he knoweth he hath but a little time." (vss. 10-12)

Wesley's commentary on this vision begins with a treatment of Michael, a created angel who fights Satan in the heavens and brings an end to his "accusing the saints." The Hebrew name of this archangel means "who is like God?" In the Epistle of Jude, Michael is depicted as not making any railing accusations against Satan but only saying, "The Lord rebuke thee" (verse 9). Moreover, Wesley stated -

> "This modesty is implied in his very name. . . which implies also his deep reverence toward God: the very name of Michael asks, 'Who is like God?' Not Satan: not the highest archangel. It is he likewise that is afterward employed to seize, bind, and imprison that proud spirit."[15]

It is this Michael who casts Satan and his wicked angels out of heaven to the earth. Wesley observed, "So till now he had a place in heaven. How deep a mystery is this! One may compare this with Luke X, 18; Eph. II, 2; IV, 8; VI, 12."[16] With the casting out of Satan, all the heavenly host rejoiced. However, he came to earth but not for the first time. Satan had visited earth before, many times before, "although his ordinary abode was in heaven."[17] Yet, after his being cast out, earth became his temporary abode.

Wesley dated the casting of Satan onto the earth as A.D. 864, so that the third woe of the Apocalypse "falls in the tenth century, extending from 900 to 1000; called the dark, the iron, the unhappy age."[18] At the beginning of this woe, Satan's wrath was poured out against the woman (Church) in the form of violent persecution. Having its genesis in the tenth century, this persecution carried over into the eleventh century -

> "In Prussia, King Adelbert was killed in the year 997, King Brunus in 1008; and when King Stephen encouraged Christianity in Hungary, he met with violent opposition. After his death, the heathens in Hungary set themselves to root it out, and prevailed for several years. About the same time, the army of the emperor, Henry III, was totally overthrown by the Vandals. These, and all the accounts of those times, show with what fury the dragon then persecuted the woman."[19]

Verse 14 of this twelfth chapter proves to be the key to Wesley's interpretation of the final signs of the end and Christ's coming. No other verse in the Bible was as pivotal for unlocking the eschatological secrets Wesley kept encountering. The text reads - "And there was given to the woman the two wings of the great eagle, that she might fly into the wilderness to her place, where she is fed for a time, and times, and half a time, from the face of the serpent." Eagles, Wesley noted, are symbols of great potentates. For instance, in Ezekiel XVII, 3, "a great eagle" refers to the king of Babylon. "Here the *great eagle* is the Roman Empire: the *two wings*, the eastern and western branches of it."[20] The "wilderness" mentioned here is not the same "wilderness" as in verse 6. There it meant Europe, from the Danube westward. Here, Wesley argued, it means both branches of the Roman Empire, so that "both emperors now lent their wings to the woman, and provided a safe abode for her."[21] Wesley moved hastily to the time dimension of verse 14, for it is this emphasis which is the most crucial to his understanding. As noted earlier in the previous chapter, the Apocalypse has a period of time identified as "non-chronos." That reappears in Wesley's commentary on Revelation XII, 14, together with other time codes - "the little time," "the time, times, and half a time," and "the time of the beast." By comparing the prophecy and history, he reasoned, these time periods "seem to begin and end nearly thus -

(1) The non-chronos extends .... from about 800 to 1836
(2) The 1,260 days of the woman .... from 847 to 1524
(3) The little time .... from 947 to 1836
(4) The time, times, and half .... from 1058 to 1836
(5) The time of the beast is between the beginning and end of the three times and a half. In the year 1058 the empires had a good understanding with each other, and both protected the woman. The bishops of Rome, likewise, particularly Victor II, were duly subordinate to the emperor."[22]

Underneath these periods of time lie the foundations of specific quantitative values, assigned to each category by some hidden logic on the part of Bengelius and Wesley. For instance, the length of "non-chronos" is 1,111 years of common time; the little time is 888 years; the time, times, and half a time is 777 years; and the time of the beast is 666 years.[23]

Verses 15 to 17 bring chapter twelve to a dramatic close -

"And the serpent cast out of his mouth after the woman water as a river, that he might cause her to be carried away by the stream. But the earth helped the woman, and opened her mouth, and swallowed up the river which the dragon had cast out of his mouth. And the dragon was wroth with the woman, and went forth to make war with the rest of her seed, who keep the commandments of God, and retain the testimony of Jesus."

In exegeting this passage, Wesley deciphered "water" as "an emblem of a great people" - in this case, the Turks, who overran the Christian part of Asia about A.D. 1060. From thence, they poured into Europe, spreading until they "had overflowed many nations."[24] The "*earth helped the woman*" refers, Wesley claimed, to the "powers" or political entities of "the earth." European powers joined together in fending off the Turks, thus saving the Church. "The time" of one of these assaults was from A.D. 1058 to 1280, "during which the Turkish flood ran higher and higher, though frequently repressed by the emperors."[25] The "two times" were from A.D. 1280 to 1725, during this period the Turkish "power flowed far and wide" but the "princes of the earth helped the woman, that she was not carried away by it."[26] The "half time" is from A.D. 1725 to 1836, and already at its beginning Turkish power has been exerted against Persia (Iran), making it difficult for her to afflict the woman in her "place." Wesley made a startling prediction - "Near the end of the 'half time,' (it will) be swallowed up, perhaps by Russia, which is risen in the room of the eastern empire."[27] And so, finally, being unable to destroy the Church in her "place," Satan goes forth to other lands to war against the scattered, "real Christians, living under heathen or Turkish governors."[28] From the beginning of chapter twelve, one sign after another has followed in progression, reaching down into the "half time" that began in 1725. Yet, Wesley is hardly finished interpreting the Apocalypse.

Revelation XIII - The vision is of a beast arising out of the sea, having seven heads, ten horns, and ten diadems upon his horns, and upon his heads a blasphemous name. Verses 2 through 18 give explicit descriptions of this fearsome, wild beast who receives the authority and power of the dragon "to make war on the saints." The significance of Wesley's exposition of this chapter lies in his identification of this beast with the Roman Papacy, with a running historical commentary. In good logical fashion, he stated his case with eight propositions -

Proposition 1 - This beast is the same beast as described in chapter XVII.

Proposition 2 - This beast "is a spiritually secular power, opposite to the kingdom of Christ." As such, he is a mixture of spiritual and secular. He is both a secular prince and a false prophet.

Proposition 3 - The beast has "a strict connexion with the city of Rome. This clearly appears from the seventeenth chapter."

Proposition 4 - "The beast is now existing." He is not past nor "altogether to come." He was, and is, and shall be until the destruction of Rome when the beast shall be thrown into the fiery lake.

Proposition 5 - "The beast is the Romanish Papacy. This manifestly follows from the third and fourth propositions. . . therefore, either there is some other power more strictly connected with that city, or the Pope is the beast."

Proposition 6 - "The Papacy, or the papal kingdom, began long ago."[29]

To advance the credibility of his logic, Wesley recited papal history -

"A.D.	1033.	Benedict IX, a child of eleven years old, is bishop of Rome.
A.D.	1048.	Damasus II introduces the use of the triple crown.
A.D.	1058.	The Church of Milan is, after long opposition, subjected to the Roman.
A.D.	1073.	Hildebrand, or Gregory VII, comes to the throne.
A.D.	1076.	He deposes and excommunicates the emperor.
A.D.	1077.	He uses him shamefully and absolves him.
A.D.	1080.	He excommunicates him again, and sends a crown to Rodulph, his competitor.

A.D. 1083. Rome is taken. Gregory flees. Clement is made Pope, and crowns the emperor.

A.D. 1085. Gregory VII dies at Salerno.

A.D. 1095. Urban II holds the first popish council, at Clermont, and gives rise to the crusades.

A.D. 1111. Paschal II quarrels furiously with the emperor.

A.D. 1123. The first western general council in the Latern. The marriage of priests is forbidden.

A.D. 1132. Innocent II declares the emperor to be the Pope's liegeman, or vassal.

A.D. 1143. The Romans set up a governor of their own, independent of Innocent II. He excommunicates them, and dies. Celestine II is, by an important innovation, chosen to the Popedom without the suffrage of the people: the right of choosing the Pope is taken from the people, and afterward from the clergy, and lodged in the cardinals.

A.D. 1152. Eugene II assumes the power of canonizing saints.

A.D. 1155. Adrian IV puts Arnold of Brixia to death for speaking against the secular power of the Papacy.

A.D. 1159. Victor IV is elected and crowned. But Alexander III conquers him and his successor.

A.D. 1168. Alexander III excommunicates the emperor, and brings him so low, that

A.D. 1177. He submits to the Pope's setting his foot on his neck.

A.D. 1204. Innocent III sets up the Inquisition against the Vaudois.

A.D. 1208. He proclaims a crusade against them.

A.D. 1300. Boniface VIII introduces the year of Jubilee.

A.D. 1305. The Pope's residence is removed to Avignon.

A.D. 1377. It is removed back to Rome.

A.D. 1378. The fifty years' schism begins.

A.D. 1449. Felix V, the last Antipope, submits to Nicholas V.

A.D. 1517. The Reformation begins.

A.D. 1527. Rome is taken and plundered.

A.D. 1557. Charles V resigns the empire; Ferdinand I thinks the being crowned by the Pope superfluous.

A.D. 1564. Pius IV confirms the Council of Trent.

A.D. 1682. Doctrines highly derogatory to the papal authority are openly taught in France.

A.D. 1713. The constitution *Unigenitas*.

A.D. 1721. Pope Gregory VII canonized anew.

...The secular princes now favored the kingdom of Christ; but the bishops of Rome vehemently opposed it. These at first were plain ministers or pastors of the Christian congregation at Rome, but by degrees they rose to an eminence of honour and power over all their brethren: till, about the time of Gregory VII (and so ever since), they assumed all the ensigns of royal majesty; yea, of a majesty and power far superior to that of all other potentates on earth."[30]

Proposition 7 - Hildebrand, or Gregory VII, is the proper founder of the papal kingdom. Patrons of the Papacy allow that he made "many considerable additions to it; and this very thing constituted the beast, by completing the spiritual kingdom: the new maxims and the new actions of Gregory all proclaim this."

To fortify this seventh proposition, Wesley quoted numerous maxims advanced by Gregory VII -

(1)   That the Bishop of Rome is universal bishop.

(2)   That he alone can depose bishops, or receive them again.

(3)   That he alone has power to make new law in the church.

(4)   That he alone ought to use the ensign of royalty.

(5)   That all princes ought to kiss his foot.

(6)   That the name of the Pope is the only name under heaven; and that his name alone should be recited in the churches.

(7)   That he has power to depose emperors.

(8)   That no general synod can be convened but by him.

(9)   That no book is canonical without his authority.

(10)   That none on earth can repeal his sentence, but he alone can repeal any sentence.

(11)   That he is subject to no human judgment.

(12)   That no power dare to pass sentence on one who appeals to the Pope.

(13)   That all weighty causes everywhere ought to be referred to him.

(14)   That the Roman church never did, nor ever can, err.

(15)   That the Roman bishop, canonically ordained, is immediately made holy, by the merits of St. Peter.

(16)   That he can absolve subjects from their allegiance."[31]

These maxims, Wesley argued, are "the genuine sayings" of Pope Gregory, according to "the most eminent Romish writers." Moreover, he continued, his actions as the bishop of Rome "agree with his words." Under a "spiritual pretext" this pope acted as if he were the "emperor of the whole Christian world." His excommunication of Henry IV of Germany, Wesley held, was an instance of this usurpation of power. The bull of excommunication read -

> "Blessed Peter, prince of the apostles, incline, I beseech thee, thine ears, and hear me thy servant. In the name of the omnipotent God, Father, Son, and Holy Ghost. I cast down the emperor Henry from all imperial and regal authority, and absolve all Christians, that were his subjects, from the oath whereby they used to swear allegiance to true kings. And moreover, because he had despised mine, yea, thy admonitions, I bind him with the bond of an anathema."[32]

Not content to depart this subject, Wesley quoted another, later pronouncement of Gregory against Henry -

> "Blessed Peter, prince of the apostles, and thou Paul, teacher of the Gentiles, incline, I beseech you, your ears to me, and graciously hear me. Henry, whom they call emperor, hath proudly lifted up his horns and his head against the church of God - who came to me, humbly imploring to be absolved from his excommunication - I restored him to communion, but not to his kingdom - neither did I allow his subjects to return to their allegiance. Several bishops and princes of Germany, taking this opportunity, in the room of Henry, justly deposed, chose Rodulph emperor, who immediately sent ambassadors to me, informing me that he should always remain in the disposal of God and us. Henry then began to be angry, and at first entreated us to hinder Rodulph from seizing his kingdom. I said I would see to whom the right belonged, and give sentence which should be preferred. Henry forbade this. Therefore I bind Henry and all his favours with the bond of an anathema, and again take from him all regal power. I absolve all Christians from their oath of allegiance, forbid them to obey Henry in anything, and command them to receive Rodulph as their king. Confirm this, therefore, by your authority, ye most holy princes of the apostles, that all may now at length know, as ye have power to bind and loose in heaven, so we have power to give and take away on earth, empires, kingdoms, principalities, and whatsoever men can have."[33]

As noted in the previous chapter, Wesley linked the "man of sin" (II Thess. II, 4), often termed "the antichrist," with a medieval pope and his successors. It is here in Wesley's commentary on the

Apocalypse (chapter XIII) that Gregory VII is identified as being the first pope in a series of popes who collectively are "the man of sin" and the "antichrist." However, in the text of Revelation XIII they are not spoken of as such, but they are called the "beast" which "arose out of the sea (Europe)." Wesley explained -

> "Thus the time of the ascent of the beast is clear. The apostasy and mystery of iniquity gradually increased till he arose 'who opposeth and exalteth himself above all (2 Thess. II, 4). Before the seventh trumpet the adversary wrought more secretly; but soon after the beginning of this, the beast openly opposes his kingdom to the kingdom of Christ."[34]

Proposition 8 - Gregory's empire "properly began in the year 1077." When the emperor, in that year, left Italy, Gregory extended his power to its maximum - "And on the first of September. . . he began his famous epoch."[35] Wesley added to this last proposition twenty-three observations, each linking Gregory and his successors with the "beast" who is also the "antichrist."

Observation 1 - "The beast is the Romish Papacy, which has now reigned for some ages."

Observation 2 - "The beast has seven heads and ten horns."

Observation 3 - "The seven heads are seven hills, and also seven kings. One of the heads could not have been, as it were, 'mortally wounded' had it been only a hill."

Observation 4 - "The ascent of the beast out of the sea is different from his ascent out of the abyss: the Revelation often mentions both the sea and the abyss; but never uses the terms promiscuously."

Observation 5 - "The heads of the beast do not begin before his rise out of the sea, but with it."

Observation 6 - "These heads, as kings, succeed one another."

Observation 7 - "The time which they take up in this succession is divided into three parts. 'Five' of the kings signified thereby are fallen; 'one is, the other is not yet come.'"

Observation 8 - "'One is;' namely, while the angel was speaking this. He placed himself and St. John in the middlemost time, that he might. . . point out the first time as past, the second as present, the third as future."

Observation 9 - "The continuance of the beast is divided in the same manner. The beast 'was, is not, will ascend out of the abyss' (Rev. XVII, 8, 11). . . . 'Five are fallen, one is, the other is not yet come.'" (a parallelism)

Observation 10 - "Babylon is Rome. All things which the Revelation says of Babylon agrees to Rome, and Rome only."

Observation 11 - "The beast reigns both before and after the reign of Babylon. First the beast reigns (Rev. XIII, 1); then Babylon (Rev. XVII, 1); and then the beast again (Rev. XVII, 8)."

Observation 12 - "The heads are the substance of the beast; the horns are not. The wound of one of the heads is called 'the wound of the beast' itself (verse 3); but the horns, or kings, receive the kingdom 'with the beast' (Rev. XVII, 12)."

Observation 13 - "The forty-two months of the beast fall within the first of the three periods. The beast rose out of the sea in the year 1077. A little after, power was given him for forty-two months. This power is still in being."

Observation 14 - "The time when the beast is 'not,' and the reign of 'Babylon' are together."[36]

Observation 15 - "The difference there is between Rome and the Pope, which has always subsisted, will then be most apparent." Here Wesley launched into a lengthy distinction between these two. The serious student will enjoy reading this passage from the *EXPLANATORY NOTES UPON THE NEW TESTAMENT* (page 1006).

Observation 16 - "In the first and second periods of his duration, the beast is a body of men; in the third, an individual. The beast with seven heads is the Papacy of many ages; the seventh head is the man of sin, antichrist. He is a body of men from Rev. XIII, 1 to XVII, 7; he is a body of men and an individual, chapter XVII, from the eighth to the eleventh verse; he is an individual from chapter XVII, 12 to XIX, 20."

Observation 17 - "That individual is the seventh head of the beast, or, the other king after the five and one, himself being the eighth, though one of the seven. As he is a Pope, he is one of the seven heads. But he is the eighth or not a head, but the beast himself. . . To illustrate this by a comparison: suppose a tree of seven branches, one of which is much larger than the rest; if those six are cut away, and the seventh remain, that is the tree."[37]

Observation 18 - "He is the wicked one, the man of sin, the son of perdition, equally termed antichrist."

Observation 19 - "The ten horns, or kings, 'receive power as kings with the wild beast one hour' (Rev. XVII, 12); with the individual beast, 'who was not.' But he receives his power again, and the kings with it, who quickly give their new power to him."

Observation 20 - "The whole power of the Roman monarchy, divided into ten kingdoms, will be conferred on the beast (Rev. XVII, 13, 16, 17)."

Observation 21 - "The ten horns and the beast will destroy the whore (verse 16)."

Observation 22 - "At length the beast, the ten horns, and the other kings of the earth, will fall in that great slaughter (chapter XIX, 19)."

Observation 23 - "Daniel's fourth beast is the Roman monarchy, from the beginning of it, till the thrones are set. This, therefore, comprises both the apocalyptic beast, and the woman, and many other things."[38]

This last proposition was important to Wesley. It linked the Old Testament apocalyptic book of Daniel with the New Testament's Apocalypse. The continuity of prophetic expectation concerning end things gave Wesley a confidence that his interpretation was not developing along purely subjective lines. Consequently, in describing the Roman monarchy that flourished under Gregory, Wesley used both Daniel's beast and Revelation's beast as one and the same - "Daniel's fourth beast is the Roman monarchy. . . This . . . comprises the apocalyptic beast."[39] But more importantly, the monarchy is like a river, running in its channel from its fountain, taking in other rivers in its course. "The Roman power was at first undivided; but it was afterwards divided into various channels, till the grand division into the eastern and western empires, which likewise underwent various changes." Whatever power the Romans had before Gregory VII, "that Daniel's beast contains; whatever power the Papacy has had from Gregory VII, this the apocalyptic beast represents."[40]

The exegesis of chapter XIII, in the *EXPLANATORY NOTES UPON THE NEW TESTAMENT*, finally gets under way, Wesley having completed his propositions and observations.

The "wild beast" is the Papacy, from A.D. 1077 onward. One of its heads had a "deadly wound" that had been healed (verse 3). Not surprisingly, Wesley ignored the rather popular *Neron redivivus* legend in interpreting this phrase. The reason for such neglect lies in Wesley's system of dating the events of Revelation. Emperor Nero (ruled A.D. 37-68) is anachronistic if added to Wesley's system. Moreover, the seven heads are, in his system, "papal kings" and not Roman emperors. Wesley's version of the deadly wound claims that the first head of the wild beast was struck by "the sword" of secular potentates who resisted him, notably the German emperors of whom were Henry III and Henry IV. Wesley observed -

"These had for a long season had the city of Rome, with her bishop, under their jurisdiction. Gregory determined to cast off this yoke from his own and to lay it on the emperor's shoulders. He broke loose, and excommunicated the emperor, who maintained his right by force, and gave the Pope such a blow, that one would have thought the beast must have been killed thereby, immediately after coming up. But he (Gregory) recovered, and grew stronger than before. The first head of the beast extends from Gregory VII, at least to Innocent III."[41]

Two deadly symptoms accompanied this wound: (1) Schisms and open ruptures ruled the church. There were five great divisions between A.D. 1080 and 1176, and a great many "antipopes;" and (2) under Pope Innocent II, the Roman nobility re-established the ancient commonwealth, stripping the Pope of secular power and "left him only his episcopal power." Popes Innocent II and Celestine II "fretted themselves to death."[42] Pope Lucius II, with sword in hand, attacked the city and died after being struck with a stone. Popes Alexander III and Lucius III were both driven from the city by an angry populace. Popes Urban III and Gregory VIII spent their days in banishment.[43]

Nevertheless, the "whole earth wondered at him." By "the whole earth" is meant "the western world," Wesley reasoned. The Pope's admirers in the west followed him "in his councils, his crusades, and his jubilees." The popes led the western peoples in the worship of the dragon, "although they knew it not." The wild beast uttered blasphemies against God and those that "dwell in heaven." The beast had the bones of many dug up, and he cursed them "with the deepest execrations." He "made war" with the saints - such as the "Waldenses and Albigenses."[44] In A.D. 1208, Pope Innocent III proclaimed a crusade against them. A great majority of them were slaughtered by the army in 1209. Wesley added, "It is Christ who shed his own blood; it is antichrist who sheds the blood of others!"[45] And yet, he continued, "his last and most cruel persecution is to come." A prophetic warning to all Christian pilgrims was thus sounded by a Wesley who could not envision the coming of a pope like John XXIII nor a council called Vatican II.

The beast's mark on the right hand of everyone under papal authority (XIII, 16) is actually the name of the beast, "namely, that of *Papa* or *Pope*.[46] The meaning of receiving this mark is -

"Whosoever, therefore, receives the mark of the beast does as much as if he said expressly, 'I acknowledge the present Papacy, as proceeding from God;' or, 'I acknowledge that what St. Gregory VII has done, according to his legend (authorized by Benedict XIII), and what has been maintained in virtue thereof, by his successors to this day, is from God.' By the former, a man hath the name of the beast as a mark; by the latter, the number of his name. In a word, to have the name of the beast is to acknowledge his holiness; to have the number of his name is to acknowledge the papal succession."[47]

Students of the Apocalypse will be amazed to see that Wesley took the beast's number, 666, and referred it to "papal succession" instead of a person's name arrived at by numerical decoding. The number of the Papacy, for Wesley, is 666 years - "So long shall he endure from his first appearing."[48] It did not fit Wesley's scheme to use a Greek manuscript of the Apocalypse in which the beast's number is given as 616 instead of 666, although he knew the alternate attested to by the Fathers.

Revelation XIV - According to Wesley, this is the last chapter in Revelation that treats things now being fulfilled. Some of these things were set in motion centuries ago, and now they are close to the divine telos. Some things are just now being set in motion, and they too move toward their consummation in Christ. At the outset of Revelation XIV, John of Ephesus sees 144,000 saints standing on mount Sion, before the Lamb of God. Wesley's exposition betrays his frequent Platonic squint - the mount is not the earthly one but the heavenly. The great number of saints have the "name" of the Lamb and the "name" of his Father written on their foreheads. The exact identity of these saints is hard to establish, he argued. Either they are those "out of all mankind who had been the most eminently holy, or the most holy out of the twelve tribes of Israel."[49] If the latter, they were those sealed by the mighty angel "to preserve them from the plagues that were to follow" (Rev. VII, 4). Whoever they may have been, they are safe now in heaven upon the eternal Mt. Sion. They form a magnificent chorus, singing with great propriety and elegance. They are the Church above, the Church Triumphant. Wesley said of them -

> "The Church above, making suitable reflections on the grand events which are foretold in this book, greatly serves to raise the attention of real Christians, and to teach the high concern they have in them. Thus is the Church on earth instructed, animated, and encouraged by the sentiments, temper, and devotion of the Church in heaven."[50]

The music of the Church above, consisting of voices and instruments, offers "a new song" which describes newly begun activities by God, much as the Enthronement Psalms did in earlier times. In the Greek text, the 144,000 saints are spoken of as *parthenoi* (virgins), "who have not defiled themselves with women." Rather than giving this statement to literal and narrow interpretation, Wesley took the broader meaning - these saints are those who have "preserved universal purity."[51] Sexual purity is not enough to please God. Total holiness, of body, mind, and soul, is required if one is to see God.[52] This understanding of purity led Wesley to the universal meaning. Moreover, these saints follow closely after the Lamb. Their reward is to be near Christ. They are "firstfruits to God" and are "without fault." Let the Church on earth learn the state of the Church in heaven, Wesley said, for in so learning there is great encouragement.

The second vision in chapter XIV features another angel, flying through heaven, with "an everlasting gospel to preach to them that dwell on the earth, and to every nation, and tribe, and tongue, and people" (verse 6). With a loud voice the angel commands - "Fear God and give glory to him; for the hour of his judgment is come: and worship him that made the heaven, and the earth, and the sea, and fountains of water!" That "the hour of his judgment is come" is a "joyful message" for Wesley - the end is nearer than at any previous time, and the end means the beginning of a new creation![53]

Immediately, in John's vision, another angel appears, announcing that "Babylon the great is fallen. . . she that hath made all nations drink of the wine of her fornication" (verse 8). Wesley observed that ancient Babylon was "magnificent, strong, proud, powerful: so is Rome also. Babylon was first, Rome afterwards, the residence of the emperors of the world." Wesley noted that what Babylon was to Israel of old, Rome is to the "spiritual Israel of God," the Church of the new age. As the liberty of the ancient Jews was linked to the destruction of Babylon, so the liberty of the Church is linked to the overthrow of Rome.[54] Also, he observed that when Babylon is mentioned in the Apocalypse, the adjective "great" always modifies it. His reasoning for this is -

". . . to teach us that Rome then commenced Babylon, when it commenced the great city; when it swallowed up the Grecian monarchy and its fragments, Syria in particular; and, in consequence of this, obtained dominion over Jerusalem about sixty years before the birth of Christ. Then it began, but it will not cease to be Babylon till it is finally destroyed. Its spiritual greatness began in the fifth century, and increased from age to age. It seems it will come to its utmost height just before its final overthrow."[55]

It should be noted that Wesley's interpretation slips from political Rome to ecclesiastical Rome without so much as a blink of an eye. Rome's "fornication" is her idolatry of venerating both saints and angels, her worship of images, her human traditions, her outward pomp, and her bloody zeal.[56] Spiritual fornication, he argued, is usually accompanied by "fleshly fornication" of which ecclesiastical Rome seemed to have an abundant supply - "Witness the stews there, licensed by the Pope, which are, no inconsiderable branch of his revenue. This is fitly compared to wine, because of its intoxicating nature." One wonders why Wesley did not quote Petrarch at this point - "Popes, little old men, chasing naked prostitutes through the palace at Avignon." But the "wine" of fornication, as Wesley understood it, was enjoyed by many who were not in the papal office. The popes distributed the wine, far and wide, and innumerable consumers partook of it to excess. "Of this wine she hath, indeed, made all nations drink," Wesley concluded, closely following verse 8.[57] Wesley saw other aspects to this wine of fornication - inquisitions, congregations, and Jesuit doctrines and practices.[58]

In John's vision, a third angel appears suddenly and warns worshipers of the wild beast that they "shall drink of the wine of the wrath of God, which is poured unmixed into the cup of his indignation" (verses 9-10). The "wine of her fornication" is here contrasted with "the wine of the wrath of God." Those who have imbibed the former shall imbibe the latter. There is no escaping this consequence. The reference to God's wine being "unmixed" has a fatalistic element - "*The wrath of God, which is poured* unmixed - Without any mixture of mercy: without hope."[59] Wesley was confident that "real anger" on the part of God results when he considers the church's wine of fornication - "And is no real anger implied in all this? Oh, what will not even wise men assert, to serve an hypothesis!"[60] The worshipers of the beast shall be tormented for ever and ever, the text says, and the smoke from their torment will ascend perpetually. Wesley, commenting on this, said - "God grant thou and I may never try the strict, literal eternity of this torment!"[61]

From the frightful scene of torment, the chapter turns dramatically to assert the blessedness of the righteous dead. John reports that he heard a voice out of heaven - "Write, From henceforth happy are the dead who die in the Lord: Yea, saith the Spirit, that they may rest from their labours. Their works follow them" (verse 13). Wesley believed that the voice may have been that of a "departed saint." "From henceforth" the righteous dead are happy, Wesley claimed, because - (1) they escape the approaching calamities; and (2) they already enjoy "so near an approach to glory."[62] The "rest" that is their's is without pain. They are not in "Purgatory" but in Paradise of Abraham's Bosom (consult Chapter IX - THE PILGRIM AND DEATH). Resting in Paradise, they enjoy "pure, unmixed happiness." Wesley exclaimed, "How different this state from that of those (verse 11) who 'have no rest day or night!' Reader, which wilt thou choose?"[63] As for their works, they do not go "before" them to secure "admittance into the mansion of joy; but they follow them when admitted."[64]

Verse 14 introduces a vision of harvest and vintage as two separate "visitations," as Wesley termed them. The text reads -

"And I looked, and behold a white cloud, and on the cloud one sitting like a son of man, having a golden crown on his head, and a sharp sickle in his hand. And another angel came out of the temple, crying with a loud voice to him that sat on the cloud, Thrust in thy sickle, and reap: for the time to reap is come; for the harvest of the earth is ripe. And he that sat on the cloud thrust in his sickle upon the earth; and the earth was reaped.

And another angel came out of the temple which is in heaven, and he also had a sharp sickle. And another angel came out from the altar, who had power over fire; and cried with a loud voice to him that had the sharp sickle, saying, Thrust in thy sickle, and lop off the clusters of the vine of the earth; for her grapes are fully ripe. And the angel thrust in his sickle upon the earth, and lopped off the vine of the earth, and cast it into the great winepress of the wrath of God. And the winepress was trodden without the city, and blood came out of the winepress, even to the horses' bridles, one thousand six hundred furlongs." (verses 14-20)

In the first visitation, "many good men are taken from the earth by the harvest." Wesley saw this as a gracious act on God's part. However, he did not expand the theme, leaving it to his readers to imagine whatever they wished. The second visitation is "penal" in nature, involving "sinners." As for the one who "is like a son of man," Wesley identified him as an angel sent by Christ, and he is not Christ himself. The golden crown he wore symbolizes his "high dignity."[65] When this angel receives word to begin his harvest, he is told the "harvest is ripe," meaning that the good men of the earth have attained a "high degree of holiness" and have "an earnest desire to be with God."[66] And the earth was reaped of its saints, but not by Christ.

In exegeting the second visitation, Wesley treated the "angel from the altar" as being associated with the heavenly altar of burnt offerings, the altar to which the martyrs had cried for vengeance (VI, 9-10). In God's economy of things, it was not the season to answer these cries. Consequently, this angel comes forth from the altar to order the immediate reaping of the wicked upon the earth. Being cast into the winepress like grapes, the wicked - ripe with fornications of many types - were "trodden" underfoot "by the Son of God (Rev. XIX, 15)." The "city" mentioned in verse 20 was, for Wesley, Jerusalem. Wesley observed that what flowed from the winepress was not wine but blood - flowing for over two hundred miles, "through the whole land of Palestine."[67]

What did this mean to Wesley? The visitation passage must stand in his scheme of last things as a transition from things now being fulfilled to things to be fulfilled shortly. Already, Wesley believed, the angel, like a son of man, is perched above the earth, sickle in hand, waiting for the order to reap, at a time known only to the Father. The same is the case concerning the reaping of the vintage. Wesley believed that his days were eschatologically transitional. Indeed, the Lord was at hand! Still the year 1836 was "not yet," and there were many things to be "shortly fulfilled" before then.

B. *Things to be Fulfilled Shortly* (Revelation XV - XIX)

Revelation XV - This brief passage of but eight verses treats the preparation for pouring seven bowls of divine wrath upon the earth. The seven angels, who take turns pouring out their bowls, are first instructed in their tasks within the temple of the "tabernacle of the testimony." Wesley called this temple "the holiest of all" because it is the heavenly house of God, eternal, not made with hands. So, the plan to afflict the earth with seven bowls of wrath originates with God and issues from his perfect tabernacle.[68] The seven angels, attired in white linen like the priests of Israel, depart the heavenly temple and begin their commission.

Revelation XVI - The angels move swiftly in performing their work. The plagues which they unleash are similar to those of Exodus, when Israel was in Egypt, except instead of ten there are but seven here.[69] The first brought "grievous" ulcers on those who "had the mark of the wild beast, and that had worshipped his image" (verse 2). The second poured out his bowl upon the sea. In consequence of this, the sea became as the blood "of a dead man" and "every living soul in the sea died." The third poured wrath onto the rivers and fountains of water, "and they became blood." With the third bowl the voice of the "angel of the waters" was heard - "righteous art thou, who art, and who wast, the Gracious one, because thou hast judged thus. For they have shed the blood of saints and prophets, and thou hast given them blood to drink: they are worthy" (verses 5-6). Another voice suddenly responded to this affirmation - a voice from the altar - "Yea, Lord God Almighty, true and righteous are thy judgments." The fourth angel spilled his bowl upon the sun, and those with the mark of the wild beast were scorched with fire. Yet they did not repent.

Wesley noted that the first four bowl plagues are connected, but the last three treat three different subjects - "The fifth concerns the throne of the beast, the sixth the Mahometans, the seventh chiefly the heathens."[70] The fifth plague falls upon the "throne" of the beast, and not on the "beast." Wesley viewed this as meaning that the papal chair will become vacant and remain so, and thus the kingdom will be darkened with "a lasting, not a transient, darkness."[71] As a result, the worshipers of the beast "*gnawed their tongues* - Out of furious impatience. Because of their pains and because of their ulcers - Now mentioned together, and in the plural number, to signify that they were greatly heightened and multiplied."[72] The sixth bowl was poured out upon the great river Euphrates. The entire river was dried up, making a way for the kings of the east to come westward and attack the Turkish empire, which is Mahometan.[73] The presbyter John reported that three "unclean spirits" went forth like frogs, having come from the mouths of the dragon, the wild beast, and the false prophet (verse 13). They are identified as "spirits of devils, working miracles." According to Wesley, the "dragon fights chiefly against God; the beast, against Christ; the false prophet, against the Spirit of truth."[74] These devils are likened to frogs because as frogs dwell in unclean places, such as marshes and fens, so these devils dwell in spiritually unclean places. They mobilize the "kings of the whole world" to come to war against God on the "great day."

The "second beast," who appears after the darkening of the kingdom of the wild beast, is also called the "false prophet" by Wesley. This is the beast of Revelation XIII, 11, which arose out of the earth. In chapter XVI, he is referred to as the "false prophet," who spawns from his mouth one of the three devil spirits. Wesley identified Mahomet with this second beast.[75] So, the dragon (Satan), the wild beast (the Papacy without a pope), and the false prophet (Mahometism) inspire the frog-devils to gather the pagans of the world to do battle with God Almighty. For Wesley this sign was about to develop and become reality.

In the midst of this scenario of the approaching Armageddon, the voice of Jesus Christ is raised (verse 15) - "Behold, I come as a thief. Happy is he that watcheth, and keepeth his garments, lest he walk naked, and they see his shame." Wesley's commentary on this affirmation of Christ is brief but adequate - "Suddenly, unexpectedly. Observe the beautiful abruptness. I - Jesus Christ. Hear him. *Happy is he that watcheth* - Looking continually for Him that 'cometh quickly.' *And keepeth his garments* - Which men use to put off when they sleep. *Lest he walk naked, and they see his shame* -

Lest he loses the graces which he takes no care to keep, and others see his sin and punishment."[76]

Concerning the reference to Armageddon (verse 16), Wesley defined the term geographically - "the mountain of Megiddo," then he cited its historical significance as an ancient battleground - "This was a place well known in ancient times for many memorable occurrences; in particular, the slaughter of the kings of Canaan, related in Judges V, 19."[77] He rightly postponed any further explanation until it reappears in chapter XIX, 19.

At last, the seventh angel pours out the seventh bowl of God's wrath upon the air. At once a loud voice from the heavenly temple proclaims, "It is done." Then came lightnings, voices, thunders, and earthquakes "such as had not been since men were upon the earth, such an earthquake, so great" (verse 18). And Jerusalem, "the great city," was split into three parts, "and the cities of the nations fell: and Babylon the great was remembered before God, to give her the cup of the wine of the fierceness of his wrath." And all the islands fled away, and there were no mountains left. Hail, the weight of a talent, fell upon men, "and the men blasphemed God, because of the plague of the hail; for the plague thereof is exceeding great" (verses 19-21). Wesley expected these things to happen in his own day. As noted earlier, for this reason, he was preoccupied with earthquakes. The seven bowl plagues, he believed, are about to begin, and once they do, the time to the coming of Christ is short indeed.

Revelation XVII - The judgment of the great whore dominates this apocalyptic passage. Wesley saw its meanings as vital for recognizing the last signs when they take place.

One of the seven angels, who earlier had a bowl of wrath, invites the prophet John to "Come hither" and witness the judgment of the "great whore that sitteth upon many waters." This notorious woman was guilty of committing fornication with the kings of the earth, and the people of the earth had gotten drunk on the "wine of her fornication" (verse 2). John consents and is carried in spirit by the angel into "a wilderness." There he beholds a woman astride a scarlet wild beast, "full of names of blasphemy," having seven heads and ten horns. As for the woman, she is clothed in purple and scarlet, wearing gold and precious stones and pearls, "having in her hand a golden cup, full of abomination and filthiness of her fornication." On her forehead a name is written - "MYSTERY, BABYLON THE GREAT, THE MOTHER OF HARLOTS AND ABOMINATIONS OF THE EARTH" (verses 3-6). Moreover, she is drunk "with the blood of the saints, and with the blood of the witnesses (martyrs) of Jesus." John, upon seeing her, wonders greatly.

Wesley's commentary on this part of John's vision is composed of simple analogies. The great whore sits as a queen, "in pomp, power, ease, and luxury." The "many waters" upon which she is first seated represent "many people and nations." The "kings of the earth" are both ancient and modern rulers. Her "fornication" is "idolatry and various wickedness." "The inhabitants of the earth" are simply "the common people." All her followers become "drunk" from the wine of her fornication, thus proving the vast influence of this whore. The "wilderness" to which John was carried is *Campagna di Rome* - "The country round about Rome, (it) is now a wilderness, compared to what it was once."[78] The term "woman" frequently means "city." "A scarlet wild beast" refers to the same "wild beast" of chapter XIII. In that chapter the "wild beast" carried on his own activity. In this passage "he is connected with the whore." The "seven heads" of this beast are "seven hills upon which the woman sitteth." The woman is the "city of Rome, with its buildings and inhabitants; especially the nobles." "Purple and scarlet" were the colors of "the imperial habit: the purple, in times of peace; and the scarlet, in times of war." The "golden cup" - like ancient Babylon (Jer. LI, 7). "Full of abominations" - "most abominable doctrines and practices." On the whore's forehead was the name "MYSTERY" - "This very word was inscribed on the front of the Pope's mitre, till some of the Reformers took public notice of it." "Babylon the great" is defined by Pope Boniface XIII in his proclamation of Jubilee (A.D. 1725) -

"To this holy city, famous for the memory of so many holy martyrs, run with religious alacrity. Hasten to the place where the Lord hath chose. Ascend to this New Jerusalem, whence the law of the Lord and the light of evangelical truth hath flowed forth into all nations, from the very first beginning of the Church; the city most rightfully called 'The Palace,' placed for the pride of all ages, the city of the Lord, the Sion of the Holy One of Israel. This catholic and apostolical Roman church is the head of the world, the mother of all believers, the faithful interpreter of God and mistress of all churches."[79]

Wesley's decoding of apocalyptic language, found in chapter XVII, continues. "The mother of harlots" means "the parent, ringleader, patroness, and nourisher of many daughters, that closely copy after her."[80] "Drunk with . . .blood" - "So that Rome may well be called 'The slaughter-house of the martyrs' - She hath shed much Christian blood in every age; but at length she is even drunk with it, at the time to which this vision refers."[81]

The angel continues his narration concerning the woman and the beast, from verse seven through verse eighteen, John being attentive. Wesley's commentary again runs along basic and simple lines. "The beast which thou sawest" - "He (1) was; (2) and is not; (3) and will ascend out of the bottomless pit, and go into perdition." Furthermore, "The seven heads are seven hills and seven kings: (1) five are fallen; (2) one is; (3) the other is not come; and when he cometh, he must continue a short space."[82] "They are seven kings" - Wesley observed that in ancient times there were royal palaces on all the seven Roman hills. "These were the Palatine, Capitoline, Coelian, Exquiline, Viminal, Quirinal, Aventine hills." But, he argued, "the prophecy respects the seven hills at the time of the beast, when Palatine was deserted and the Vatican in use. Not that the seven heads mean hills distinct from kings; but they have a compound meaning, implying both together." Wesley then speculated -

"Perhaps the first head of the beast is the Coelian hill, and on it the Latern, with Gregory VII and his successors; the second, the Vatican, with the church of St. Peter, chosen by Boniface VIII; the third, the Quirinal, with the church of St. Mark, and the Quirinal palace built by Paul II; and the fourth, the

Exquiline hill, with the temple of St. Maria Maggiore, where Paul V reigned. The fifth will be added hereafter."[83]

To avoid criticism for this bit of speculation, Wesley added another argument -

"In the papal register, four periods are observable since Gregory VII. In the first almost all bulls made in the city are dated in the Latern; in the second, at St. Peter's; in the third, at St. Mark's or in the Quirinal; in the fourth, at St. Maria Maggiore. But no fifth, sixth, or seventh hill has yet been the residence of any Pope. Not that one hill was deserted, when another was made the papal residence; but a new one was added to the other sacred palaces."[84]

Leaving the subject of the seven hills that are also seven kings (popes with secular as well as spiritual authority), Wesley gave an interesting time sequence of signs of the end -

A.D.	1058.	Wings are given to the woman.
A.D.	1077.	The beast ascends out of the sea.
A.D.	1143.	The forty-two months begin.
A.D.	1810.	The forty-two months end.
A.D.	1832.	The beast ascends out of the bottomless pit.
A.D.	1836.	The beast in finally overthrown.[85]

As for the "ten horns" of the beast, Wesley followed the text which claims them to be "ten kings." Wesley was scrupulously correct when he asserted that -

"It is nowhere said that these horns are on the beast, or on his heads. And he is said to have them, not that he is one of the seven, but he is the eighth. They are ten secular potentates, contemporary with, not succeeding, each other, who receive authority as kings with the beast, probably in some convention, which, after a very short space, they will deliver up to the beast. Because of their short continuance, only authority as kings, not a kingdom, is ascribed to them. While they retain this authority together with the beast, he will be stronger than ever before; but far stronger still, when their power is also transferred to him."[86]

Wesley noted that these ten kings, as the text indicates, shall make war with the Lamb, and that the Lamb will overcome them, for he is "Lord of lords, and King of kings."[87]

Ironically, Wesley recognized, the whore shall become an object of great hatred, hated fiercely by the ten kings and the beast, who shall "make her desolate and naked, and shall eat her flesh, and burn her with fire" (verse 16). The reason given for this unexpected turning upon the woman is equally startling - "For God hath put in their hearts to execute his sentence, and to agree, and to give their kingdom to the wild beast, till the words of God shall be fulfilled" (verse 17).

Revelation XVIII - "Babylon the great is fallen, is fallen, and is become an habitation of devils, and an hold of every unclean spirit, and a cage of every unclean and hateful bird," cried the angel from heaven, as his luminous glory lightened the earth (verses 1-2).

Wesley, habitually prolific with words and concepts, does little in his commentary to illumine the reader about this particular vision. "Babylon is fallen - This fall was mentioned before (Rev. XIV, 8); but is now declared at large." "And is become an habitation" simply means "a free abode." "Of devils, and an hold" refers to "a prison." "Of every unclean spirit" implies "perhaps confined there where they had once practiced all uncleanness, till the judgment of the great day." Wesley's most lucid statement reads -

> "How many horrid inhabitants hath desolate Babylon! of invisible beings, devils, and unclean spirits; of visible, every unclean beast, every filthy and hateful bird. Suppose then, Babylon to mean heathen Rome; what have the Romans gained, seeing from the time of that destruction, which they say is past, these are to be its only inhabitants for ever?"[88]

The text of Revelation resumes the flow of prophecy - Another voice out of heaven is heard by the prophet. Wesley claimed it to be the voice of Christ, saying,

> "Come out of her, my people, that ye may not be partakers of her sins, and that ye receive not of her plagues. For her sins have reached even to heaven, and God hath remembered her iniquities. Reward her even as she hath re-

warded, and give her double according to her works: in the cup which she mingled, mingle to her double. As much as she hath glorified herself, and lived deliciously, so much torment and sorrow give her: because she saith in her heart, I sit as a queen, and am no widow, and shall see no sorrow. Therefore shall her plagues come in one day, death, and sorrow, and famine; and she shall be burned with fire: for strong is the Lord God who judgeth her. And the kings of the earth, who had committed fornication and lived deliciously with her, shall weep and mourn over her, when they see the smoke of her burning. Standing afar off for fear of her torment, saying, Alas, alas, thou great city Babylon, thou strong city! In one hour is thy judgment come." (verses 4-10)

Again, Wesley's exposition of the passage is scant. In a scattered fashion he treated bits and pieces of the text, as if to say, "The words of Christ are sufficient. Let them speak for themselves." Some of his better comments include - "*Reward her* - This God speaks to the executioners of His vengeance." "*As she rewarded* - Others; in particular, the saints of God." "*Give her double* - This, according to the Hebrew idiom, implies only a full retaliation." "*Lived deliciously* - In all kinds of elegance, luxury, and wantonness." "*I sit* - Her usual style. Hence those expressions, 'The chair, the see of Rome: he sat so many years.'" "*As a queen* - Over many kings, 'mistress of all churches; the supreme; the infallible; the only spouse of Christ; out of which there is no salvation.'" "*Thou strong city* - Rome was anciently termed by its inhabitants, Valencia, that is, strong." Wesley added to this the Greek word for Rome which "signifies strength."[89]

After a brief interlude of narrative in the text, in which the reaction of merchants to the fall is stated, the pronouncement of Christ against the whore is resumed (verses 14-20). In this dirge the merchants stand afar off from Rome and weep and mourn. And why not? Their former business of "any thing for a denarius" - made possible by the whore - is no more. They cry, "In one hour so great riches are become desolate." Then ship-masters and sailors, standing afar off, raise their lamentation for the proud but fallen city - "What city was like the great city? . . .wherein were made rich all that had ships. . . in one hour she is made desolate." The dirge is suddenly terminated by Christ, commanding the hosts of heaven, the saints, apostles, and prophets to "rejoice over her. . . for God hath avenged you on her" (verse 20).

The chapter concludes with a mighty angel throwing a millstone into the sea, signifying how the great city shall be cast down and "found no more at all" (verses 21ff). Rome, once proud of its fine arts - music and visual arts - shall be deprived of such sophistications. Instead of music, only the crashing sound of the falling millstone shall be heard. Not even the light of a single candle shall be seen in her. There shall be no marriages in her after she is fallen, hence no festive music and celebration, and, consequently, no children. Fallen and desolate, indeed. The last verse (verse 24) - "And in her was found the blood of prophets, and saints, and all that had been slain upon the earth" - gave Wesley the occasion to wax eloquent in his commentary -

"The same angel speaks still, yet he does not say 'in thee,' but in her, now so sunk as not to hear these last words. *And of all that had been slain* - Even before she had been built. See Matthew XXIII, 35. There is no city under the sun which has so clear a title to catholic blood-guiltiness as Rome. The guilt of the blood shed under the heathen emperors has not been removed under the Popes, but hugely multiplied. Nor is Rome accountable only for that which hath been shed in the city, but for that shed in all the earth. For at Rome under the Pope, as well as under the heathen emperors, were the bloody orders and edicts given: and wherever the blood of holy men was shed, there were the grand rejoicings for it. And what immense quantities of blood have been shed by her agents! Charles IX of France, in his letter to Gregory XIII, boasts, that in and not long after the massacre of Paris he had destroyed seventy thousand Huguenots. Some have computed that, from the year 1518 to 1548, fifteen millions of Protestants have perished by the Inquisition. This may be overcharged; but certainly the number of them in those thirty years, as well as since, is almost incredible. To these we may add innumerable martyrs, in ancient, middle, and late ages, in Bohemia, Germany, Holland, France, England, Ireland, and many other parts of Europe, Afric, and Asia."[90]

Revelation XIX - At long last, the end is in sight! This graphic chapter unfolds the divine activities that immediately precede the consummation of all things. There are three parts to this chapter - (1) verses 1-8, depicting the joy in heaven because God the Almighty truly reigns; (2) verses 9-10, stating the blessedness of those invited to the holy marriage; and (3) verses 11-21, revealing the coming in triumph of Christ as King of kings and Lord of lords.

(1) The combined voices of the heavenly multitude sounded forth as one "loud voice," praising God -

"Hallelujah: the salvation, and the glory, and the power of our God: For true and righteous are his judgments: for he hath judged the great whore, who corrupted the earth with her fornication, and hath avenged the blood of his servants at her hand." (verses 1-2)

This act of praise was followed immediately by another, involving the twenty-four elders and the four living creatures, earlier mentioned in Revelation IV. Their praise was short but potent - "Amen; Hallelujah" (verse 4). Another voice spoke, from the throne - an indication that this heavenly scene is identical to that of chapter IV. This speaker commanded universal worship of God - "Praise our God, all ye servants, and ye that fear him, small and great" (verse 5). The obedient response was overwhelming - "I heard. . . a voice of a great multitude, and as a voice of many waters, and as a voice of mighty thunders." The corporate voice proclaimed -

"Hallelujah; for the Lord God, the Almighty reigneth. Let us be glad and rejoice, and give glory to him: for the marriage of the Lamb is come, and his wife hath made herself ready. And it is given to her to be arrayed in fine linen, white and clean; the fine linen is the righteousness of the saints" (verses 6-8).

What did this mean to Wesley? The passage was very important to him, as his commentary implies. For instance, every verse is treated in the *EXPLANATORY NOTES UPON THE NEW TESTAMENT*, except verse three, which is redundant. The majority of these verse-treatments are lengthy, quite unlike his exegesis in preceding chapters. An exploration of his commentary, verse by verse, for this particular passage, will prove beneficial.

Verse 1 - "*. . .a loud voice of great multitude* - Whose blood the great whore had shed." "*Saying, Hallelujah* - This Hebrew word signifies, *Praise ye Jah*, or *Him that is*. God named himself to Moses, EHELEH, that is, *I will be* (Exod. III, 14)." Wesley continued the explanation of the divine name, noting that God gave Moses his name as "Jehovah, that is, 'He that is, and was, and is to come.'" Moreover, in Revelation XVI, 5, God is styled, "He that is and was" and not "he that is to come." In completing this discussion, Wesley affirmed -

> "At length He is styled, 'Jah.' 'He that is:' the past together with the future being swallowed up in the present, the former things being no more mentioned, for the greatness of those that now are. This title is of all others the most peculiar to the everlasting God."[91]

"*Salvation*" is shown by Wesley to be deliverance from "the destruction which the whore had brought upon the earth." As for God's "*power and glory*," they appear from his "judgment executed on her, and from the setting up His kingdom to endure through all ages."[92]

Verse 2 - "*For true and righteous are his judgments* - This is the cry of the souls under the altar changed into a song of praise."[93]

Verse 4 - "*The four and twenty elders and the four living creatures* . . .

> - The living creatures are nearer the throne than the elders. Accordingly, they are mentioned before them, with the praise they render to God (Rev. IV, 9-10; V, 8, 14); inasmuch as there the praise moves from the centre to the circumference. But here, when God's judgments are fulfilled, it moves back from the circumference to the centre. Here, therefore, the four and twenty elders are mentioned before the living creatures."[94]

Verse 5 - The voice which came from the throne, Wesley believed, was that of one of the four living creatures. This call to praise is to commemorate the fact that "God, the Almighty, takes the kingdom to Himself, and avenges Himself on the rest of His enemies." Wesley asked a pertinent question - "Were all these inhabitants of heaven mistaken? If not, there is real, yea, and terrible anger in God."[95]

Verse 6 - The voice of the great multitude of all God's servants is lifted in praise for the "*Almighty reigneth* - More eminently and gloriously than ever before."[96]

Verse 7 - As for the "*marriage of the Lamb*," Wesley was mystified as to its precise meaning, while he boldly affirmed that it is "near at hand, to be solemnized speedily." Of its precise meaning he said - "What this implies, none of 'the spirits of just men,' even in paradise, yet know."[97]

Verse 8 - The "*bride*" is clearly "all holy men, the whole invisible Church." To be arrayed in "*fine linen*" is an emblem of the righteousness of the saints - "Both of their justification and sanctification."[98]

(2) The blessedness of being invited to the marriage supper of the Lamb (verses 9-10) is predicated upon being called "to glory" and responding in faith, with obedience and holiness of heart and life.[99]

(3) Verses 11 through 21 "show the magnificent expedition of Christ and His attendants against His great adversary."[100] When he comes, as here described, he shall ride "forth on his white horse, with the sword of his mouth." In the time of his humiliation, Christ rode on a lowly ass. But his parousia shall be with triumph. So, the white horse is his emblem of a victorious hero appearing in "solemn triumph."[101] This conquering Christ is called "Faithful and True" - by performing all his promises and in executing all his threatenings.[102] His flaming eyes attest to his omniscience, and his many diadems make him the King of all nations.[103] His actual name is known only to himself - "As God He is incomprehensible to every creature." His clothing is "dipped in blood" - the blood of his enemies. "He hath already conquered." He shall rule those whom he conquers, with a rod of iron, "that is, if they will not submit to His golden sceptre." And he executes his judgments on the ungodly, by treading the winepress of God's wrath. The ruler of all nations,

once a child, now appears as "a victorious warrior." Wesley expanded this theme in a forceful manner -

"The nations have long ago felt His 'iron rod,' partly while the heathen Romans, after their savage persecution of the Christians, themselves groaned under numberless plagues and calamities, by His righteous vengeance; partly while other heathens have been broken in pieces by those who bore the Christian name. For although the cruelty, for example, of the Spaniards in America was unrighteous and detestable, yet did God therein execute His righteous judgment on the unbelieving nations; but they shall experience His iron rod as they never did yet, and then will they all return to their faithful Lord."[104]

Consequently, this Christ is rightly heralded as KING OF KINGS, AND LORD OF LORDS. He earns this title by virtue of all that he has done and what he is about to accomplish on the field of battle against his enemies. As he comes from heaven, he brings with him his "armies" on white horses. However, his chief weapon is a sharp two-edged sword, coming from his mouth, with which he shall "smite the nations." Standing in the sun, an angel calls the birds of prey to come to the battle and gorge themselves on the flesh of Christ's enemies and mounts. This shall be called "the great supper of God."[105] Wesley saw the ten kings (Revelation XVII, 12) gathering other kings and their armies to this battle with Christ. This is the famous Battle of Armageddon that popular interpretation distorts. Wesley did not view Armageddon as a great world war with nation pitted against nation, the end result being the utter reduction to chaos. He found the meaning of Armageddon in the text of Revelation XIX - all the kings of the earth come together to oppose Christ. As quickly as the battle begins, it ends. Christ is victor. The enemies become the food of scavengers. The wild beast and the false prophet are then cast "alive" into the lake burning with brimstone -

"The beast and the false prophet plunge at once into the estremest degree of torment, without being reserved in chains of darkness till the judgment of the great day. Surely none but the beast of Rome would have hardened himself thus against the God he pretended to adore, or refused to have repented under such dreadful, repeated visitations! Well is he styled a beast, from his carnal and vile affections; a wild beast from his savage and cruel spirit!"[106]

Here endeth the recitation of final signs of the end and of Christ's coming in triumph, as Wesley found them in the Apocalypse. However, Wesley acknowledged other signs, signs not specifically revealed in the Bible, but nevertheless signs to be recognized as God's faithful fulfilling of all things in Christ.

THE FINAL SIGNS IN THE EIGHTEENTH CENTURY

Wesley often spoke to his Methodist pilgrims about the signs of the very last days. Those predicted in II Timothy III, 1 - 7, appeared as early as the time of St. Paul, but they seemed to have a renaissance in Wesley's time. As always, they were of a negative nature - Men will be lovers of themselves; they will be lovers of money; they will be arrogant and proud, evilspeakers, and disobedient to parents; they will be known as being ungrateful and unholy; their affection will be unnatural; being incapable of pacification, they will be slandereres, intemperate, fierce, and despisers of good men; they will be traitors, rash and puffed up with their own importance; they will be lovers of pleasure (*aphilagathoi* - Greek, "not lovers of good") more than lovers of God; they will have a form of godliness, but not the power of it; they are those who are silly with women, being led about by their unstable desires; and they are always learning something, but they never arrive at any truth. The eighteenth century was surely afflicted with these signs of the last days, the Church even more so. For this reason, Wesley believed that divine judgment must begin first with the house of God.[107]

However, God has already begun to judge the Church, with justice and mercy, renewing and purifying it, pouring out his Spirit upon it to fulfill its evangelical mission. This is also a sign that the end is near, and a very positive sign! Wesley, in holding this position, often spoke of Methodism as being a valuable part of this work of God. In a letter, dated June 13, 1771, Wesley wrote - "It is plain to me, that the whole work of God termed Methodism is an extraordinary dispensation of his providence."[108] "An extraordinary dispensation" - Indeed! A sign that the end of time is near! In his famous sermon, THE SIGNS OF THE TIMES, Wesley affirmed this belief -

> "The times which we have reason to believe are at hand are what many pious men have termed the time of the 'latter-day glory;' meaning the time wherein God would gloriously display his power and love in the fulfillment of his gracious promise, that 'the knowledge of the Lord shall cover the earth, as the waters cover the sea.'"[109]

Methodism, as the sign of God's "latter-day glory," was designed by God to "propagate Bible religion through the land; that is, faith working by love; holy tempers and holy lives."[110] The land, primarily, meant England, although many other lands became recipients of this divine design. In his *A FURTHER APPEAL TO MEN OF REASON AND RELIGION*, Wesley argued -

> "God is now visiting this nation, in a far other manner than we had cause to expect. . . Instead of pouring out his fierce displeasure upon us, he has made us yet another tender mercy; So that even when sin did most abound, grace hath much more abounded."[111]

In fact, he continued, salvation, as understood and experienced by Methodist pilgrims, has abounded so rapidly and extensively that "our fathers" never could have imagined such a thing. "When hath religion, I will not say since the Reformation, but since the time of Constantine the Great, made so large a progress in any nation, within so small a space?" Wesley added, "I believe, hardly can either ancient or modern history supply us with a parallel instance."[112] Moreover, Methodism is not "a slight or superficial thing." The spiritual fruits of holiness are worthy of the apostolic age.[113] Hence the movement has the remarkable quality of purity - purity because its doctrine is the biblical theology of the Church of England;[114] purity because it is rational as well as scriptural;[115] pu-

rity because it is entirely free from bigotry;[116] purity because "this
religion has no mixture of vice or unholiness;"[117] and, lastly, purity
because "the spirit of persecution" is completely absent from the
Methodists.[118]

While Methodism was a "new sign" of the kingdom of God
nearing its consummation, Wesley insisted that it was not a "new"
religion. In his sermon *AT THE FOUNDATION OF CITY ROAD
CHAPEL*, he made this clear -

> "What is Methodism? What does this new word mean? Is it not a new reli-
> gion? This is a very common, nay, almost an universal, supposition; but
> nothing can be more remote from the truth. It is a mistake all over.
> Methodism, so called, is the old religion, the religion of the Bible, the religion
> of the primitive Church, the religion of the Church of England. This old reli-
> gion. . . is 'no other than love, the love of God and all mankind; the loving
> God with all our heart, and soul, and strength, as having first loved us, - as
> the fountain of all good we have received, and of all we ever hope to enjoy;
> and the loving every soul which God hath made, every man on earth as our
> own soul. This love is the great medicine of life; the neverfailing remedy for
> all the evils of a disorderly world; for all the miseries and vices of men.
> Wherever this is, there are virtue and happiness going hand in hand; there is
> humbleness of mind, gentleness, long-suffering, the whole image of God;
> and, at the same time, a 'peace that passeth all understanding,' with 'joy un-
> speakable and full of glory.' This religion of love, and joy, and peace, has its
> seat in the inmost soul; but is ever showing itself by its fruits, continually
> springing up, not only in all innocence, but, likewise, in every kind of benefi-
> cence - spreading virtue and happiness to all around it."[119]

Believing that God raised up Methodism to extend divine love
and mercy prior to the end of time, Wesley included the story of
the rise of Methodism in his sermon entitled *THE WISDOM OF
GOD'S COUNSELS*. Wesley frequently spoke of the beginning
and development of Methodism, but here he placed it within the
context of God's eternal wisdom and plan. Its origin is in eternity *a
parte ante*, its program is very late in time, and its destiny is eternity
a parte post.

Tracing Methodism's rise in time, Wesley began the story in
A.D. 1725 -

". . .about the year 1725. . . . Mr. Law published his 'Practical Treatise on Christian Perfection,' and, not long after, his 'Serious Call to a Devout and Holy Life.' Here the seed was sown, which soon grew up, and spread to Oxford, London, Bristol, Leeds, York, and, within a few years, to the greatest part of England, Scotland, and Ireland."[120]

Wesley called Law's influence a "seed." Using the metaphor of the mustard seed (Mark IV, 30-34), Wesley's recitation emphasized the miraculous advance of growth and influence -

"March 24 (1785) - I was considering how strangely the grain of mustard seed, planted about fifty years ago, has grown up. It has spread through all Great Britain and Ireland; the Isle of Wight and the Isle of Man; then to America from the Leeward Islands, through the whole continent, into Canada and Newfoundland. And the societies, in all these parts, walk by one rule, knowing religion is holy tempers; and striving to worship God, not in form only, but likewise 'in spirit and in truth.'"[121]

The mustard seed of Methodism was sown in 1725, but it was not until 1729 that Methodism began at Oxford University as the Holy Club. In that year the seed germinated and took root in fertile ground. In his sermon ON FAMILY RELIGION, Wesley explicitly stated that the revival termed Methodism began in 1729.[122] This is significant because it roots the spiritual life of Methodism in a context other than the Aldersgate experience of May 24, 1738. Back to Oxford in 1729 -

"Let us observe what God has done already. Between fifty and sixty years ago, God raised up a few young men, in the University of Oxford, to testify those grand truths, which were then little attended to: - That without holiness no man shall see the Lord; - that this holiness is the work of God, who worketh in us both to will and to do; - that he doeth it of his own good pleasure, merely for the merits of Christ; - that this holiness is the mind that was in Christ; enabling us to walk as he also walked; - that no man can be thus sanctified till he is justified; - and that we are justified by faith alone. These great truths they declared on all occasions, in private and public; having no design but to promote the glory of God, and no desire but to save souls from death. From Oxford, where it first appeared, the little leaven spread wider and wider. More and more saw the truth as it is in Jesus, and received it in the love thereof. More and more found 'redemption through the blood of Jesus, even the forgiveness of sins.' They were born again of His Spirit, and filled with righteousness, and peace, and joy in the Holy Ghost. It afterward spread to every part of the land, and a little one became a thousand."[123]

In the City Road Chapel sermon, the more important stages of this rise were carefully traced by Wesley. In October 1735, John and Charles Wesley were joined by Mr. Ingham "to go over to the new colony in Georgia." Their design was "to preach to the Indian nations bordering on that province; but we were detained at Savannah and Frederica, by the importunity of the people, who, having no other Ministers, earnestly requested that we would not leave them." As time passed in Savannah, "the most serious of them" met with Wesley "once or twice a week" at his house - "Here were the rudiments of a Methodist society." These gatherings were not contrary in any way to the Church of England - "Both my brother and I were vehemently attached to the Church as ever." Wesley returned to England in February 1738, referring only in retrospect to his over-zealous High Church ways while in Georgia. Upon his return, he "was now in haste to retire to Oxford," and bury himself in what he called "my beloved obscurity." His account omits Aldersgate altogether, and he even gave other reasons for his preaching ministry by which he became excluded from many churches -

"I was detained in London, week after week, by the Trustees for the Colony of Georgia. In the mean time, I was continually importuned to preach in one and another church; and that not only morning, afternoon, and night, on Sunday, but on week-days also. As I was lately come from a far country, vast multitudes flocked together; but in a short time, partly because of those unwieldy crowds, partly because of my unfashionable doctrine, I was excluded from one and another church, and at length, shut out of all! Not daring to be silent, after a short struggle between honour and conscience, I made a virtue of necessity, and preached in the middle of Moorfields. Here were thousands upon thousands, abundantly more than any church could contain; and numbers among them who never went to any church or place of public worship at all. More and more of them were cut to the heart, and came to me all in tears, inquiring, with the utmost eagerness, what they must do to be saved. I said, 'If all of you will meet on Thursday evening, I will advise you as well as I can.' The first evening about twelve persons came; the next week, thirty or forty. When they were increased to about an hundred, I took down their names and places of abode, intending, as often as it was convenient, to call upon them at their own houses. Thus, without any previous plan or design, began the Methodist society in England, - a company of people associating together, to help each other to work out their own salvation."[124]

The next spring (1739), Methodist societies were formed in Bristol and Kingswood, and soon increased in other places.[125] What did "to work out their salvation" mean? Participation in a Methodist society produced certain stages of spiritual growth and development -

"Many of these (participants) were in a short time deeply convinced of the number and heinousness of their sins. They were also made thoroughly sensible of those tempers which are justly hateful to God and man, and of their utter ignorance of God, and entire inability, either to know, love, or serve him. At the same time, they saw in the strongest light the insignificancy of their outside religion; nay, and often confessed it before God, as the most abominable hypocrisy. Thus did they sink deeper and deeper into that repentance, which must ever precede faith in the Son of God. And from hence sprung 'fruits meet for repentance.' The drunkard commenced sober and temperate; the whoremonger abstained from adultery and fornication; the unjust from oppression and wrong. He that had been accustomed to curse and swear for many years, now swore no more. The sluggard began to work with his hands, that he might eat his own bread. The miser learned to deal his bread to the hungry, and to cover the naked with a garment. Indeed, the whole form of their life was changed: They had left off doing evil, and learned to do well. But this was not all. Over and above the outward change, they began to experience inward religion. 'The love of God was shed abroad in their hearts,' which they continue to enjoy to this day."[126]

Wesley was asked, "Are there now any signs that the day of God's power is approaching?" His answer was direct, and, of course, it related to the spiritual impact of Methodism -

"I appeal to every candid, unprejudiced person, whether we may or not, at this day, discern all those signs (understanding the words in a spiritual sense) to which our Lord referred John's disciples? 'The blind receive their sight:'

Those who were blind from their birth, unable to see their own deplorable
state, and much more to see God, and the remedy he has prepared for them
in the Son of his love, now see themselves, yea, and 'the light of the glory of
God in the face of Christ Jesus. . .' 'The deaf hear:' Those that were before
utterly deaf to all the outward and inward calls of God, now hear, not only his
providential calls, but also the whispers of his grace. . . 'The lame walk:'
those who never before arose from the earth or moved one step toward
heaven, are now walking in all the ways of God; yea, running 'the race that is
set before them.' 'The lepers are cleansed:' The deadly leprosy of sin, which
they brought with them into the world, and which no art of man could ever
cure, is now clean departed from them. . . At this day the gospel leaven, faith
working by love, - inward and outward holiness, - hath so spread in various
parts of Europe, particularly in England, Scotland, Ireland, in the Islands, in
the North and South, from Georgia to New-England, and Newfoundland,
that sinners have been truly converted to God, thoroughly changed both in
heart and life; not by tens, or by hundreds only, but by thousands, yea, by
myriads! The fact cannot be denied."[127]

Another sign that the end is near, according to John Wesley, is
Satan's reaction to Methodism. In his sermon *ON GOD'S VINE-
YARD*, Wesley treated the various activities of Satan in countering
the divine work of "the old religion" called Methodism. It was not
possible, he argued, that all these remarkable works of God should
be done "without a flood of opposition." Wesley's use of "flood"
was deliberate, referring to the dragon's flood against Mother
Church (Revelation XII). From the beginning of Methodism, "the
prince of this world," who was not dead, nor asleep, waged a "fight,
that his kingdom not be delivered up."[128] He surely did at Corn-
wall[129] and many other places where Methodist preaching and societies came. Satan did not come alone. He led "all his hosts to
war." At first, he stirred up mobs of people against preachers and
societies. Wesley called these mobs "beasts of the people." "They
roared like lions; they encompassed the little and defenseless on
every side." The King himself gave orders to magistrates everywhere to quell the madness of the people. As in Revelation XII,
where the princes come to the aid of the woman being accosted by
the beast, so the King alleviated Wesley's movement. Wesley
added an ironic touch to this victory over the raging dragon -

"It was about the same time that a great man applied personally to His Majesty, begging that he would please to 'take a course to stop these run-about Preachers.' His Majesty, looking sternly upon him, answered without ceremony, like a King, 'I tell you, while I sit on the throne, no man shall be persecuted for conscience' sake."[130]

Another attack upon the Methodists came in Kent. Several commissioners brought charges against a preacher and members of his society for breaking the Conventicle Act (passed by Parliament in A.D. 1664). A fine was levied in the local court, but the defendants appealed to the Court of the King's Bench and the previous action was rescinded, removing Methodist activities from under the heavy hand of the Conventicle Act.[131]

The numerous devices of Satan, to diminish and still Methodism by the perversion of English law, Wesley believed, have all failed. Such a phenomenon gave him another argument for his movement being a unique sign of the end -

"I believe this is a thing wholly without precedent. I find no other instance of it, in any age of the Church, from the day of Pentecost to this day. Every opinion, right and wrong, has been tolerated, almost in every age and nation. Every mode of worship has been tolerated, however superstitious or absurd. But I do not know that true, vital, scriptural religion was ever tolerated before. For this the people called Methodists have abundant reason to praise God. In their favour he hath wrought a new thing in the earth: He hath stilled the enemy and the avenger."[132]

But, what Satan could not do by legal means, he attempted by other means. Rather than fighting the movement from the outside, Satan, through numerous devices, began to tear it down from within. Most vulnerable to these devices were the unsaved who were seeking salvation. It must be remembered that admission into the societies was predicated on "a *desire* to flee from the wrath to come, to be saved" from one's sins, and not on already "having been saved."[133] Hence, society meetings were frequently battlegrounds on which Christ and Satan fought for the same soul. This struggle is frequently mentioned throughout Wesley's *JOURNAL*. Satan

possesses the soul of the sinner, as part of his kingdom, at the beginning of the society meeting. As Christ, through enabling grace
mediated by the Holy Spirit, calls the soul to repentance and faith,
Satan often "tears" the soul, sometimes throwing the body into fits
or convulsions.[134] Wesley reported one such incident to which he
had not been a witness, but he surely believed the report or he
would not have given it so much prominence in the *JOURNAL* -

> "In her fits she was first convulsed all over, seeming in an agony of pain, and
> screaming terribly. Then she began cursing, swearing, and blaspheming in
> the most horrid manner. Then she burst into vehement fits of laughter, then
> sunk down dead. All this time she was quite senseless; then she fetched a
> deep sigh, and recovered her sense and understanding, but was so weak that
> she could not speak to be heard, unless you put your ear almost close to her
> mouth. . . .We began singing a hymn, and quickly found His Spirit was in the
> midst of us; but the more earnestly we prayed, the more violently the enemy
> raged. . . She laid fast hold on Molly Loftis and me with inexpressible eager-
> ness; and soon burst into a flood of tears, crying, 'Lord, save me, or I perish!
> I will believe. Lord, give me power to believe; help my unbelief!' . . .I now
> looked at my watch and told her, 'It is half-hour past two: this is the time
> when the devil said he would come for you.' But, blessed by God, instead of
> a tormentor, He sent a comforter. Jesus appeared to her soul, and rebuked
> the enemy."[135]

The first occasion of fits, witnessed by Wesley, was at the society
in Bristol, in August 1739. He was surprised by it, not being aware
of such a thing ever having happened in the history of the Christian
faith. At first he thought it was entirely of the Holy Spirit. In a letter to Reverend Ralph Erskine (A.D. 1685-1752), written from
Bristol on August 24, 1739, Wesley asked this Scotch Presbyterian
evangelical if he knew of the phenomenon and whether it had been
experienced in his ministry. The letter reads as follows -

> "A great and effectual door is opened among us, and the many adversaries
> cannot shut it. But what a little surprised us at first was the outward manner
> wherein most of these were affected, who are cut to the heart by the sword of
> the Spirit. Some of them drop down as dead, having no strength nor appear-
> ance of life left in them. Some burst out into strong cries and tears, some ex-

ceedingly tremble and quake; from some great drops of sweat fall to the ground, others struggle as in the agonies of death, so that four or five strong men can hardly restrain a weak woman or child from hurting themselves or others. Of these many are in that hour filled with peace and joy; others continue days or weeks in heaviness, so that sometimes their bodies almost sink under the weight of the wounded spirit."[136]

Shortly after Wesley's letter to Erskine, he received an unsolicited one from the Reverend Joshua Read (died an old man in 1746), a Presbyterian minister at Bradford. The Reverend Mr. Read had heard of Wesley and his work at Bristol. He informed Wesley that "fits" had been around for some time. An older ministerial friend of his, a Mr. Davis, preached often as his parishioners were in the throes of fits. In fact, Mrs. Davis longed to have the experience. Davis continued his sermonizing, in spite of the fits, advising sober hearers to "Regard them not, 'twill be well enough with them, etc."[137] However, Joshua Read did offer Wesley some practical ways of evaluating the phenomenon, keeping in mind that not all who experience it are afterwards real Christians. Read began with three general observations - (1) That "much of nature" is to be seen in these fits. A "moving discourse" (sermon) has a natural tendency to raise the emotions, and "women more generally are sooner and easier affected then men." (2) There is "sometimes a diabolical agency in raising or promoting such fits." And (3) The Holy Spirit's work is to "convince of sin." The convictions of the Spirit are either common or special -

"Many come under a common work who never prove sincere converts, and yet God's ordinary way of working true grace in the souls of the elect is first by bringing 'em under a spirit of bondage by the law before he lovingly reveals his Christ and his covenant to their souls by the spirit of adoption. Thus the law is a schoolmaster to bring us to Christ, who is the end of the law for righteousness to everyone that believeth."[138]

There are four appropriate queries, Read suggested, that help in the evaluation of fits -

"First, how was it with them before they had these fits upon them? Were they before this persons of ignorance or knowledge? Were they of sober lives, or vicious and profane?

Secondly, how is it with them under those fits? Can they, after the fit is over, tell you what most affected 'em in hearing the word? Was it any particular doctrine you was upon, or come they upon any when you are treating on different and various subjects?

Thirdly, how is it with 'em after the fit is over? Are they subject to these fits only while they are hearing of you, or is it afterwards as well as then?

And finally, what can their conversations (daily life-style) witness for them?"[139]

Wesley responded to Rev. Read's letter in October 1739, agreeing with his general observations - "I believe nature might have a part in those fits, as well as Satan, raging before he is cast out; but that the Holy Spirit, deeply convincing them of sin, is the chief agent in most of those who are seized with them."[140] Wesley, using Read's questions in reference to members of the Bristol society that had experienced fits, reported that (1) some of these persons "were very ignorant before that time, some were very knowing." Some were "glossly vicious and profane; others, as touching the righteousness of the law, blameless!" (2) Afterwards, some could give "a distinct account" of the words that affected them or some single sentence, "often taken from the Holy Scripture, which suddenly pierced their soul like a dart." In that moment "they lost all command of themselves." The subjects that affected them were various, "but always bordering upon the love of Christ to lost sinners." (3) The fit usually ends as suddenly as it begins - "Either they see Christ by the eye of faith; or a scriptural promise is applied to their soul as they are filled with peace and joy, they know not how." In that instant, they regain strength in their bodies, and comfort is imparted to their souls. "Some are thus affected only once, and that while they are hearing the Word." Others, are affected many times, "chiefly in hearing; but sometimes in their own houses."

Wesley's concluding comment shows some degree of uncertainty about the effectiveness of fits - "We have reason to hope, most of those who have been seized but ONCE are indeed new creatures. Of some that were ill many times I stand in doubt, lest they should draw back into perdition."[141]

The phenomenon of fits passed from young Methodism after the Everton Revival of 1759.[142] Wesley argued that such "outward symptoms" appear at the beginning of a general work of God, as in New England, Scotland, Holland, Ireland, and many parts of England, but then they "decrease, and the work goes on more quietly and silently."[143] In evaluating the various extraordinary happenings of the Everton Revival, Wesley concluded that while much was the work of the Holy Spirit, nevertheless, Satan "mimicked this work of God, in order to discredit the whole work."[144] Indeed, in time, Methodist "fits" were scattered throughout the land as a thing of derision. Wesley spent much time and effort to correct his literate critics. In his treatise *PRINCIPLES OF A METHODIST FARTHER EXPLAINED*, he attempted to answer numerous objections raised on the subject. The serious student should consult this treatise (*WORKS*, VIII, pages 454ff) to see the scope of the controversy and how it tarnished the mature Methodism of the latter part of the eighteenth century. Wesley, in 1771, finally concluded that the Holy Spirit does not "show itself in groaning and sighings, in fits and starts." Forgetting his earlier interpretation of fits, he said, "I never affirmed it did: And when you represent me as so doing, you are a sinner against God, and me, and your own soul."[145] Fits proved, in the realm of practicality, to be a phenomenon that Satan could use against Wesley and his precious Methodists.

Clearly for John Wesley, the eighteenth century had sufficient signs of its own that Christ's coming in triumph was at hand. both Bible and history, rightly understood, consistently affirm that the telos of God is upon us. Though he died in 1791, before his eschatological expectations were realized, he could affirm with his final breath, "The best of all, God is with us" - With us in the "now" and in the "not yet" which is *almost NOW*.

Notes

1 WORKS, *op.cit.*, XII, 395.
2 Williams, *op.cit.*, 192.
3 EXPLANATORY NOTES NEW TESTAMENT, *op.cit.*, 814.
4 Williams, *op.cit.*, 194.
5 EXPLANATORY NOTES NEW TESTAMENT, *op.cit.*, 954.
6 *Ibid*, 993. Cf. WORKS, *op.cit.*, XIV, 161. In his COMPENDIUM OF LOGIC, Wesley defines logic as "the art of reasoning." See also WORKS X, 483, 491-492, in which Wesley claims that all clergymen should master it.
7 *Ibid*, 994.
8 *Loc.cit.*
9 *Loc.cit.*
10 *Loc.cit.*
11 *Ibid*, 995.
12 *Loc.cit.*
13 *Loc.cit.*
14 *Loc.cit.*
15 *Loc.cit.*
16 *Ibid*, 996.
17 *Loc.cit.*
18 *Ibid*, 998.
19 *Loc.cit.*
20 *Loc.cit.*
21 *Ibid*, 999.
22 *Loc.cit.*
23 *Loc.cit.*
24 *Ibid*, 1000.
25 *Loc.cit.*
26 *Loc.cit.* Cf. WORKS, *op.cit.*, XII, 296. Wesley had another interpretation for the "flood" from the dragon - "Bishop Browne thought Arianism and Socinianism were the flood which the dragon is in this age pouring out of his mouth to swallow up the woman... the main flood in England seems to be Antinomianism."
27 *Loc.cit.*
28 *Loc.cit.*
29 *Ibid*, 1001.
30 *Ibid*, 1001-1002.
31 *Ibid*, 1003.
32 *Loc.cit.*
33 *Ibid*, 1003-1004.

34   *Ibid*, 1004.
35   *Loc.cit.*
36   *Ibid*, 1005.
37   *Ibid*, 1006.
38   *Ibid*, 1007. Cf. EXPLANATORY NOTES OLD TESTAMENT, *op.cit.*,
       III, 2448. Wesley makes the same point here as in his twenty-third observa-
       tion.
39   *Loc.cit.*
40   *Loc.cit.*
41   *Ibid*, 1008.
42   *Loc.cit.*
43   *Loc.cit.*
44   *Ibid*, 1009.
45   *Ibid*, 1011.
46   *Ibid*, 1012.
47   *Loc.cit.*
48   *Loc.cit.*
49   *Loc.cit.*
50   *Loc.cit.* Cf. JOURNAL, *op.cit.*, VII, 150. In 1786 Wesley preached a num-
       ber of times on this text. See also, JOURNAL, VII, 445, for Wesley's All
       Saints Day sermon on this text.
51   *Ibid*, 1013.
52   WORKS, *op.cit.*, VII, 314ff.
53   EXPLANATORY NOTES NEW TESTAMENT, *op.cit.*, 1014.
54   *Loc.cit.*
55   *Loc.cit.*
56   *Loc.cit.*
57   *Loc.cit.*
58   *Ibid*, 1014-1015.
59   *Ibid*, 1015.
60   *Loc.cit.*
61   *Loc.cit.*
62   *Loc.cit.*
63   *Ibid*, 1015-1016.
64   *Ibid*, 1016.
65   *Loc.cit.*
66   *Loc.cit.*
67   *Ibid*, 1017.
68   *Ibid*, 1018.
69   *Ibid*, 1019.
70   *Ibid*, 1021.
71   *Loc.cit.*
72   *Loc.cit.*
73   *Loc.cit.*
74   *Loc.cit.*
75   *Ibid*, 1022.

76   *Loc.cit.*
77   *Loc.cit.*
78   *Ibid*, 1023.
79   *Ibid*, 1024.
80   *Loc.cit.*
81   *Ibid*, 1025.
82   *Ibid*, 1025-1026.
83   *Ibid*, 1026.
84   *Loc.cit.*
85   *Loc.cit.*
86   *Ibid*, 1027.
87   *Loc.cit.*
88   *Ibid*, 1028.
89   *Ibid*, 1029-1030.
90   *Ibid*, 1032.
91   *Loc.cit.*
92   *Ibid*, 1033.
93   *Loc.cit.*
94   *Loc.cit.*
95   *Loc.cit.*
96   *Loc.cit.*
97   *Loc.cit.*
98   *Ibid*, 1034.
99   *Loc.cit.*
100 *Loc.cit.*
101 *Loc.cit.*
102 *Loc.cit.*
103 *Loc.cit.*
104 *Ibid*, 1035.
105 *Loc.cit.*
106 *Ibid*, 1036.
107 *Ibid*, 885.
108 WORKS, *op.cit.*, XII, 356.
109 *Ibid*, VI, 307.
110 *Ibid*, XII, 428.
111 *Ibid*, VIII, 204.
112 *Ibid*, VIII, 205.
113 *Loc.cit.*
114 *Loc.cit.*
115 *Ibid*, VIII, 206.
116 *Loc.cit.*
117 *Ibid*, VIII, 207.
118 *Loc.cit.*
119 *Ibid*, VII, 423-424.
120 *Ibid*, VI, 330.
121 JOURNAL, *op.cit.*, VII, 59.

122 WORKS, *op.cit.*, VII, 77.
123 *Ibid*, VI, 281.
124 *Ibid*, VII, 422-423.
125 *Ibid*, VII, 423.
126 *Ibid*, VIII, 203-204.
127 *Ibid*, VI, 307-308.
128 *Ibid*, VII, 209.
129 JOURNAL, *op.cit.*, VII, 109.
130 WORKS, *op.cit.*, VII, 210.
131 *Loc.cit.*
132 *Loc.cit.*
133 *Ibid*, VIII, 270.
134 JOURNAL, *op.cit.*, V, 32-35.
135 *Ibid*, V, 34-35.
136 Baker, *op.cit.*, XXV, 680.
137 *Ibid*, XXV, 683.
138 *Ibid*, XXV, 684.
139 *Loc.cit.*
140 *Ibid*, XXV, 695.
141 *Ibid*, XXV, 696.
142 JOURNAL, *op.cit.*, IV, 359.
143 *Ibid*, IV, 347.
144 *Ibid*, IV, 359.
145 WORKS, *op.cit.*, IX, 189.

Chapter Eleven

The End and the Beginning:
The Coming of Christ

"Thou Lamb of God, thou Prince of peace,
　For thee my thirsty soul doth pine;
　My longing heart implores thy grace;
　O make me in thy likeness shine.

When pain o'er my weak flesh prevails,
　With lamb-like patience arm my breast;
　When grief my wounded soul assails,
　In lowly meekness may I rest.

Close by thy side still may I keep,
　Howe'er life's various currents flow;
　With steadfast eye mark every step,
　And follow thee where'er thou go.

Thou, Lord, the dreadful fight hast won;
　Alone thou hast the wine-press trod;
　In me thy strengthening grace be shown;
　O may I conquer through thy blood.

So, when on Zion thou shalt stand,
　And all heaven's hosts adore their King,
　Shall I be found at thy right hand,
　And, free from pain, thy glories sing."[1]

This German Pietistic hymn, translated by John Wesley for use throughout his Methodist societies, sought to make the coming of Christ a personal event as well as a cosmic one. The Methodist pilgrim should love the "appearing of Christ" for numerous reasons -

for the triumph of Christ over Satan, the beast, and the false prophet; for the radical and complete destruction of evil; for the restoration and improvement of the Creation; for holiness, both outward and inward, as the chosen life of every moral creature; for living in an eternal and perfect society, with God central and personally accessible; but especially for each former pilgrim to see God, face to face, and live forever in a progressively beatific state.

In that remarkable sermon, *HUMAN LIFE A DREAM*, Wesley drove home this lesson - That time and eternity meet in the coming of Christ. He boldly exclaimed,

> "Vanish then this world of shadows;
> Pass the former things away!
> Lord, appear! appear to glad us
> With the dawn of endless day!
> O conclude this mortal story,
> Throw this universe aside!
> Come, eternal King of glory,
> Now descend, and take thy bride!"[2]

Further, the individual pilgrim was advised by Wesley to "connect the ideas of time and eternity."[3] Such a connection must be more than an intellectual apprehension of concepts. To be truly a *Christian connection*, the mind and soul of a pilgrim must take a leap of faith, casting the total being of the pilgrim from time into eternity - an existential act of believing in order to know that there is a personal dimension to eternity, just as there has been a personal dimension to time. One must take the leap now, in time, and, when Christ comes, faith will become knowledge in the boundless duration of eternity.

THE COMING: I COME QUICKLY

The last words of Jesus in the Apocalypse, and Bible for that matter, are, "Yea, I come quickly" (Revelation XXII, 20). The prophet John responds enthusiastically, as Wesley believed every true pilgrim should, saying, "Come, Lord Jesus!"[4]

The *parousia* of Christ, in the Apocalypse, is set in the dramatic episode of Armageddon (Revelation XIX, 17-21). As we have seen in the previous chapter, that event is brief and rapidly over. Jesus is the mighty victor! The wild beast and the false prophet, with their worshippers, are immediately "taken" and "cast alive" into the fiery lake of eternal torment. Satan is soon captured and incarcerated in a bottomless pit (Revelation XX, 1-3). Christ has returned, in this context, and has prevailed over his enemies.

However, this particular scenario of Christ's coming is unique, found only here in all of Scripture. It is highly suggestive in its imagery, and short on details. Such a state of affairs did not seem to bother Wesley one jot or tittle. He believed that prophecy-already-fulfilled contains particulars, while prophecy-yet-to-be-accomplished exists only in figurative pictures. When the mysterious advent prophesied finally occurs, then, and only then, will mystery give way to details.

Wesley saw other "figurative" pictures of Christ's coming, in addition to the Armageddon one of a mighty conqueror. Two other important figures of the New Testament tradition, for Wesley, are: (1) Christ coming as the bridegroom, and (2) Christ coming as the thief.

(1) Christ coming as the Bridegroom. Wesley observed that the analogy of Christ as a bridegroom and his Church as his bride permeates the New Testament. He also understood this motif in the cultural context of biblical marriage. In biblical marriage, he noted, there are two stages - the legal (*erusin* in Hebrew), in which a marriage contract has been entered into and a bride price (*mohar* in Hebrew) has been agreed upon - when witnessed publicly, the bridegroom and bride are recognized as legally husband and wife; and the spiritual (*nissuin* in Hebrew), in which the relationship is consummated by participating in the holiest of all human acts - sexual intercourse between husband and wife - in which God "makes the two as one." Cohabitation follows the second stage, while the time between the erusin and nissuin is spent in courtship,

although bridegroom and bride are legally husband and wife. For more on this subject, consult Chapter VIII of this work, THE PILGRIM AND MARRIAGE.

Based upon passages in Matthew, Mark, Luke, and John, Jesus Christ was frequently termed "the bridegroom."[5] In Ephesians V, 25, Wesley noted, "Christ gave up himself" as a "model of conjugal affection" - his *mohar* - out of love for his wife, the Church.[6] In Matthew's parable of the ten virgins, Wesley explained, the night in question is the time of the *nissuin* when the husband, after a considerable courtship, comes to his wife's home to claim her and take her to a new home, thus beginning cohabitation by consummating the marriage spiritually. In the parable, the ten virgins are actually "bridesmaids" -

> "The bridesmaids, on the wedding night, were wont to go to the house where the bride was, with burning lamps or torches in their hands, to wait for the bridegroom's coming. When he drew near, they went to meet him with their lamps, and to conduct him to the bride."[7]

Five of these bridesmaids were wise and had sufficient oil for their lamps, but five were foolish and "took not oil with them." While they were out buying oil, the bridegroom came. The wise bridesmaids went off to the marriage with the bridegroom, but the foolish were late and the door was shut, locking them out in the darkness of night (Matthew XXV, 1-13). Wesley clearly understood the bridegroom to be Christ himself, having finished his courtship, coming at midnight ("an hour quite unthought of"), to take his bride, consummate the marriage, and begin the wedding feast.[8] While the bride is the Church, all individual believers are likened in the parable to either "wise" or "foolish" bridesmaids. Wesley admonished all pilgrim bridesmaids - "He that watches has not only a burning lamp, but likewise oil (love in their hearts) in his vessel. And even when he sleepeth his heart waketh."[9] The words of Jesus, concluding the parable, are given in the form of a commandment - "Watch therefore: for ye know not the day nor the hour" (verse 13). The "ignorance" theme of not knowing the "day nor hour" of the coming of Christ is sounded only here in the gospel eschatology. While the phrase "the day nor hour" appears in Matthew XXIV, 36, and Mark XIII, 32, Wesley argued, it specifically refers to "the day of judgement" in them and not to the coming of the bridegroom.[10] Nevertheless, great watchfulness is required by all pilgrims in reference to the Bridegroom's coming - No one presently knows the "day nor hour," not even John Wesley. True, he followed Bengelius in identifying the year as 1836, but soon he was criticized everywhere for that. Ironically, he had given one of his preachers, George Bell, in 1762, both private and public rebuke for teaching that "the world would end on the 28th of February."[11] On the day before, February 27, Bell and his followers "ascended a mound near the site of St. Luke Hospital." There he was arrested, "taken before a magistrate, and committed to prison, to await his predicted end of the world."[12] On the 28th, however, Wesley preached at Spitalfields on "Prepare to meet thy God!" His congregation was certainly awake. "I largely showed the utter absurdity of the supposition that the world was to end that night," Wesley recorded in his *JOURNAL*.[13] Nowhere is there a description of George Bell's "morning after" experience. Robert Southey, however, portrayed Bell as a fanatic to his dying day, turning from religion to politics.[14] All this caught up with Wesley, in 1788, when he was forced to back away from the 1836 date he had written into his *EXPLANATORY NOTES UPON THE NEW TESTAMENT* while decoding the Apocalypse. Apparently, he oftimes quoted this view in sermons, as a letter to his friend, Christopher Hopper, indicates. Wesley wrote -

I said nothing, less or more, in Bradford church, concerning the end of the world, neither concerning my own opinion, but what follows: - That Bengelius had given it as his opinion, not that the world would then end, but that the millennial reign of Christ would begin in the year 1836. I have no opinion at all upon the head: I can determine nothing at all about it. These calculations are far above, out of my sight. I have only one thing to do, - to save my soul, and those that hear me."[15]

(2) Christ coming as the thief. Matthew, the first gospel writer for Wesley, immortalized the Lord's words on this theme -

> "Watch therefore: for ye know not what hour your Lord cometh. But ye know this, that if the householder had known in what watch the thief would have come, he would have watched, and not suffered his house to be broken open. Therefore be ye also ready; for at an hour ye think not the Son of man cometh." (Matthew XXIV, 42-44)

Wesley's commentary on the passage is scant, probably because he viewed the saying of Jesus to be self-evident, full of apparent meaning, and incapable of misinterpretation.[16] He could have engaged in some linguistics, however, without diminishing the saying. For instance, in the Greek text of Matthew, the word for "thief" is *kleptas*. The English word "kleptomania" (a persistent tendency or impulse to steal) is a derivative of this Greek word. Such a thief has a tendency to "take by surprise," or to use great cleverness and cunning in accomplishing his mission. This is the linguistic thrust of all the New Testament passages in which the coming of Christ is likened to the coming of a thief. In fact, Wesley treated the term in this way in his commentary on II Peter III, 10 - "*as a thief* - Suddenly, unexpectedly."[17] Another linguistic point is pertinent here, although Wesley did not offer it. A *kleptas* thief has some degree of finesse to his work. The Greek New Testament has another term for thief - *lastas*, denoting some degree of violence in performing this act. In contrast, the *kleptas* does not engage in violence. Applied to Christ, and his coming, *kleptas* is symbolically more appropriate, and quite deliberate as used by apostolic authors. In Matthew XXIV, 43, Luke XII, 39f, Revelation II, 3, Revelation XVI, 15, Christ refers to himself as "a kleptas." However, the apostles Paul and Peter treat the matter differently. Paul, in I Thessalonians V, 3 and 5, refers to the day of Christ's coming as the "thief" and not Christ himself.[18] Peter, in II Peter III, 10, follows suit, saying, "But the day of the Lord will come as a thief."[19] The idea that Christ may be likened to a thief may have been repugnant to these apostles. Nevertheless, the coming of Christ in power and glory, will take all by surprise, the pilgrims as well as the non-pilgrims. At least pilgrims, though surprised, will be prepared for his coming.

Traditional studies of Wesley's eschatology generally have relied upon his sermon *THE GREAT ASSIZE*. We too shall examine that sermon, but not at first. The sermon is an excellent source for seeing the implications of the great judgement, but there are other important eschatological themes to be treated before that of the judgement. For now, there are several more aspects of the coming of Christ that require attention.

The manner of that advent, for Wesley, basically was related by St. Paul in I Thessalonians IV, 16f - "The Lord himself shall descend from heaven with a shout, with the voice of an archangel, and with the trumpet of God. . . in clouds. . . in the air." Wesley regarded the voice of the archangel (Michael) to be louder than the voice of Christ which is heard first, and the "trumpet" (as the voice of God) to be louder and more arresting than both the other voices. Three voices herald the coming of Christ - his own, the archangel's, and then God's.[20] Each succeeding voice is louder than its predecessor's. The appearance suddenly occurs in the clouds and air of the lower heaven. Wesley linked this Pauline statement with the Johannine affirmation of Revelation I, 7 - "Behold, he cometh with clouds; and every eye shall see him, and they who have pierced him: and all the tribes of the earth shall wail because of him." Wesley maintained that this affirmation - "he cometh" - always refers to Christ's coming in triumph, as contrasted to his first advent in which he came forth in humiliation.[21] The appearance of Christ is completely arresting because he is clothed in the glory of God - "For the Son of man shall come in the glory of the Father with his angels; and then shall he render to every man according to his work" (Matthew XVI, 27-28). Wesley claimed that this glorious theophany will be so captivating that "There is no way to escape the righteous judgment of God."[22] After all, judgment will take place when the *Christus Victor* has completed certain preliminary activities. Later, when he begins to judge, he will share that task - "When the Son of man shall sit on the throne of his glory, ye also shall sit upon twelve thrones, judging the twelve tribes of Israel" (Matthew XIX, 28). Wesley commented on this shared mission of judgment - "In the beginning of the judgment they shall stand (2 Cor. V, 10). Then being absolved, they shall sit with the Judge (1 Cor. VI, 2)."[23] Moreover, as the Son of man is still descending, lightning-like flashes attend his coming downward (Matthew XXIV, 27) - to which Wesley added, "The next coming of Christ will be as quick as lightning; so that there will not be time for any such previous warning."[24] The activities of the returning Christ, as depicted in the Apocalypse, are - (1) put down enemies at Armageddon; (2) cast the wild beast and the false prophet into the lake of fire that burns for ever and ever; (3) have Satan bound in the bottomless pit; and (4) inaugurate the "first resurrection" and the millennial reign.[25]

One of the great exegetical puzzles of New Testament eschatology is how to reconcile Paul's statement concerning what happens at Christ's coming and the first and second resurrections of the Apocalypse. Wesley was well aware of this problem, but he made no effort to solve it other than to treat Paul's position as a general statement and John's as a specific and more detailed statement of the order of things. He also knew that St. Augustine had claimed that the first resurrection is of the soul, and the second is of the body at the general resurrection. The Bishop of Hippo, in his famous *CIVITAS DEI*, had devoted an entire chapter to this subject, and Wesley knew it. In the Apocalypse, upon the coming of Christ, there is an immediate resurrection, called, in Revelation XX, 5, "the first resurrection." The righteous dead who participate in this revivification are identified as those -

"who had been beheaded for the testimony of Jesus, and for the word of God, and those who had not worshipped the wild beast nor his image, neither had received the mark on their forehead, or on their hand; and they lived and reigned with Christ a thousand years. The rest of the dead lived not again till the thousand years were ended. This is the first resurrection. Happy and holy is he that hath a part in the first resurrection: over these the second death hath no power, but they shall be priests of God and of Christ, and shall reign with him a thousand years."[26]

The Apocalypse, Wesley noticed, advances the idea of another resurrection, much later in time, involving the remainder of the righteous dead and all the other dead, small and great, after which the divine judgment begins.[27] In his sermon THE GREAT AS-SIZE, Wesley skillfully blended Paul and John together, without calling attention to the puzzle. He was justified in this, for after all, he was preaching a sermon on a great truth, a truth that does not need technical accuracy to the inth degree. Besides, in the sermon he was treating future prophetic events that point to fulfillment, but events that will only be understood with clarity and precision after they have been fulfilled.

THE RESURRECTION

Whether one prefers Paul's account of Christ's coming and its order, or John's, or Wesley's blending them together into an understandable and preachable narrative, the nature of the resurrection must be defined. The New Testament, in numerous places, affirms that at Christ's coming in glory and power the dead shall be raised and divine judgment will follow. When St. Paul preached this prophetic expectation to the philosophers at Athens, his theme of resurrection was met with contempt and rejection.[28] A century later, a converted philosopher, by the name of Athenagoras, wrote in defense of the Christian doctrine of resurrection. Contained in the volumes of the *ANTE-NICENE FATHERS*, this apology remains one of the great classics of doctrine. In it Athenagoras answered all the philosophical objections to how decomposed bodies can ever be rebuilt, especially after having been eaten by wild beasts, fish, and birds. Also, how can human flesh that is absorbed as nourishment into other beings live again? Ashes, being scattered by winds, and washed away by rains, can never be re-assembled. These and similar arguments by heathens were cleverly refuted by the apologist. Centuries passed, and similar objections to the resurrection surfaced. The time was the early eighteenth century, and the place was Christian England. A new apologist arose to attempt a refutation of neo-paganism. Dr. Benjamin Calamy,

Vicar of St. Lawrence in London, was this modern Athenagoras. Using the arguments of the earlier apologist, Calamy published his work as a sermon in 1704. The sermon influenced many, even a young student at Oxford University by the name of John Wesley. In 1732, Wesley rewrote the sermon, preaching it frequently over the years, and finally included it in his collection of sermons, under the sermon number CXXXVII and the title *ON THE RESUR-RECTION OF THE DEAD*. The sermon, following Athenagoras and Calamy, is a remarkable synthesis of Scripture and Christian reason. A study of it will reveal what Wesley believed about the resurrection.

The text of the sermon is familiar, taken from the resurrection chapter of the New Testament, I Corinthians XV, verse 35 - "But some man will say, How are the dead raised up? and with what body do they come?" The sermon begins with reference to Christ's resurrection - "It cannot now any longer seem impossible to you that God should raise the dead; since you have so plain an example of it in our Lord, who was dead and is alive."[29] The same power which raised him "must also be able to quicken our mortal bodies." This was a faith affirmation, but, for Wesley, it was also a presupposition of logic. As such, logic demands, if one mortal body can be raised from the dead, then all mortal bodies likewise may be raised by the same power. In order to logically reject the arguments of the sermon, one must reject this basic premise. In doing so, one must also reject the empirical evidence for Christ's resurrection. For Athenagoras, Calamy, and Wesley, no Christian can deny Christ's resurrection and remain a Christian.[30] His resurrection is essential to the work of salvation, and it is essential for our consummation in the eternal plan.

The sermon proceeds by stringing together a number of rational questions, posed to embarrass resurrection affirmations, and thought by some persons to be devastating -

"How are the dead raised up? And with what body do they come? How can these things be? How is it possible that these bodies should be raised again, and joined to their several souls, which many thousands of years ago were either buried in the earth, or swallowed up in the sea, or devoured by fire? - which have mouldered into the finest dust, - that dust scattered over the face of the earth, dispersed as far as the heavens are wide; - nay, which has undergone ten thousand changes, has fattened the earth, become the food of other creatures, and these again the food of other men? How is it possible that all these little parts, which made up the body of Abraham, should be again ranged together, and, unmixed with the dust of other bodies, be all placed in the same order and posture that they were before, so as to make up the very self-same body which his soul at his death forsook?"[31]

Wesley's answers to these questions do not come immediately. He felt obliged first to give an argument used by critics against resurrection, drawn from Ezekiel XXXVII. The passage treats the prophet's vision of a valley filled with dry bones. A noise was heard and a shaking, "and the bones came together, bone to bone." Sinews and flesh came upon them, and skin covered them. Then breath entered into them. Suddenly they lived "and stood upon their feet." The critics argued that this was only a vision of a deranged prophet, nothing more -

"That our bones, after they are crumbled into dust, should really become living men; that all the little parts whereof our bodies were made, should immediately, at a general summons, meet again, and every one challenge and possess its own place, till at last the whole be perfectly rebuilt; that this, I say, should be done, is so incredible a thing, that we cannot so much as have any notion of it."[32]

But, Wesley held, such a principle of reunited matter is possible because created matter is eternal and indestructible. It is reasonable to apply this principle to disorganized, mortal bodies. In this line of argument, Wesley was following the christianized atomism of Democritus (B.C. 460-360), sanitized by Athenagoras in his apology, and reintroduced by Benjamin Calamy.

There are three main sections to the sermon on resurrection, each defining the parameters of the subject. (1) That the resurrection of the body that died and was buried is not incredible or impossible; (2) The scriptural differences between the qualities of a glorified and a mortal body; and (3) The inferences to be drawn from this whole matter.

(1) Resurrection: Incredible or Impossible? In approaching this subject logically, the "plain notion" of a resurrection requires that "the self-same body that died should rise again." Nothing "can be said to be raised again, but that very body that died." If, for instance, God were to give new bodies to our souls, at the last day, "this cannot be called the resurrection of our body; because that word plainly implies the fresh production of what was before." Wesley turned to Scripture - "St. Paul. . . tells us that this corrupt-ible must put on incorruption, and this mortal must put on immor-tality." Both the "corruptible" and the "mortal" refer to our present bodies "which we now carry about with us, and one day shall lay down in the dust."[33] Daniel predicted that "those who sleep in the dust of the earth shall awake; some to everlasting life, and some to everlasting contempt." These terms, "sleep" and "awake," imply that when "we rise again from the dead, our bodies will be as much the same as they were when we awake from sleep." Furthermore, Wesley continued, if the same body does not rise again, why should God open all the graves at the end of the world? "The graves can give up no bodies but those which were laid in them."[34] Moreover, St. Paul claims that "the Lord shall change this vile body, that it may be fashioned like unto his glorious body." Consequently, this "vile body" can be no other than that which we are "now clothed, which must be restored to live again."[35] There is nothing incredible or impossible about this, Wesley argued. God is able to "distinguish and keep unmixed from all other bodies the particular dust into which our several bodies are dissolved, and can gather it together and join it again, how far soever dispersed assunder."[36] Since God knows all the stars of heaven by name, and is able to number the sands of the sea-shores, it should be a small thing for him to know our individual particles of dust that compose our bodies - even af-ter these particles have been dispersed to death and decomposi-tion. How strange it would be, if he who formed us in our mothers' wombs, could not recognize every particle that belongs to us. "The artist knows every part of the watch which he frames," Wesley

quoted from Calamy. Moreover, the watch-maker marks every part he has made, with his own mark, and he knows the use of each part. How much more does God our Father, who has made all our parts, putting his mark upon each particle, know our parts and their usage, displaying an infinite knowledge of his works, far surpassing the finite knowledge of the watch-maker.[37] "Why may not the same power (that created us) collect the ruins of our corrupted bodies, and restore them to their former condition?"[38] Indeed, Wesley continued, God's power is not limited -

"All the parts into which men's bodies are dissolved, however they seem to us carelessly scattered over the face of the earth, are yet carefully laid up by God's wise disposal till the day of restoration of all things. They are preserved in the waters and fires, in the birds and beasts, till the last trumpet shall summon them to their former habitation."[39]

As for human flesh that was eaten by wild beasts, birds, and fish, being transformed as nourishment into the flesh of lower creatures - How can these particles ever be restored? Quite easily, Wesley claimed, since only "a very small part of what is eaten turns into nourishment, the far greater part goes away according to the order of nature."[40] Question answered!

The God who made man from the dust of the earth (Genesis II, 7), can likewise "re-make" his body although death has dissolved its organized constitution. Man's creation and man's resurrection are equally marvelous, employing the same power of accomplishment, the same wisdom, and the same result - a living body. When God has raised this body -

"He can enliven it with the same soul that inhabited it before. And this we cannot pretend to say is impossible to be done; for it has been done already. Our Saviour himself was dead, rose again, and appeared alive to his disciples and others, who had lived with him many years, and were then fully convinced that he was the same person they had seen die upon the cross."[41]

Wesley, following Calamy's sermon, turned the attack around, demanding that the critics of resurrection explain various appearances in nature, if they can. When they are able to explain every phenomenon of nature, then let "them talk of the difficulties of explaining the resurrection."[42] For instance, let them explain how the first drop of blood was made. "How came the heart, and veins, and arteries to receive it?" "Of what, and by what means, were the nerves and fibres made?"[43] But, if they cannot answer these things without "having recourse to the infinite power and wisdom of the FIRST CAUSE, let them know that the same power and wisdom can re-animate it, after it has turned into dust."[44]

(2) The scriptural qualities of mortal and glorified bodies.

Wesley believed that there are four qualitative differences between a mortal body and a resurrected one. First, the body "we shall have at the resurrection" shall be *immortal* and *incorruptible*. Wesley, ever quotable, should speak to this -

"Now these words not only signify that we shall die no more, (for in that sense the damned are immortal and incorruptible,) but that we shall be perfectly free from all the bodily evils which sin brought into the world; that our bodies shall not be subject to sickness, or pain, or any other inconveniences we are daily exposed to."[45]

Scripture calls this "the redemption of our bodies," that is, "the freeing them from all their maladies."[46] Any wise man would choose this redemption over letting his body "lie rotting in the grave."[47] Compared to this marvelous body is the mortal body that precedes it. Wesley claimed that the best that can be said about the mortal body is that it is "a house of earth," and "a ruinous building," always ready to "tumble into dust."[48] As such, it is not our home. We look for another "house, eternal in the heavens," a resurrected and glorified body. Until then, our mortal bodies are frail at best. "How soon are they disordered! To what a troop of diseases, pains, and other infirmities are they constantly subject!" Moreover, Wesley continued, the "least distemper" of the mortal body disturbs the mind, making life a burden. When one of the many bodily parts is afflicted adversely, "the whole man suffers."[49] Furthermore, the sustaining of the mortal body in health requires great and constant effort, not always successful. But, Wesley exclaimed, "When we shall have passed from death unto life, we shall be eased of all the troublesome care of our bodies, which now takes up so much of our time and thoughts."[50] Expanding upon this hope, Wesley added -

"We shall be set free from all those mean and tiresome labours which we must now undergo to support our lives. Yon robes of light, with which we shall be clothed at the resurrection of the just, will not stand in need of those careful provisions which it is so troublesome to us here either to procure or to be without. But then, as our Lord tells us, those who shall be accounted worthy to obtain that world 'neither marry nor are given in marriage, neither can they die any more, but they are equal to the angels.' Their bodies are neither subject to disease, nor want that daily sustenance which these mortal bodies cannot be without. . . This is that perfect happiness which all good men shall enjoy in the other world, - a mind free from all trouble and guilt, in a body free from all pains and diseases. Thus our mortal bodies shall be raised immortal."[51]

Secondly, our bodies shall be raised in glory. Wesley defined "glory" in terms of light or effulgent brightness. As noted earlier in Chapter II (GOD'S ATTRIBUTES AND THE TRINITY), God enjoys this property as expressive of his essential nature. For Wesley, the biblical resurrection of the righteous means the manifestation of glory in the risen and transformed body. Stating it in other terms, the resurrected body shares the visible nature of God's being. Glory, therefore, becomes the property of one's essential being, one that has been raised immortal and incorruptible, as Christ was first raised. This personal aura of light, Wesley suggested, is similar to the "lustre of Moses's face, when he had conversed with God on the mount."[52] The face of Stephen, the first Christian martyr in Acts VII, shone radiantly with that same glory before the Council, "as it had been the face of an angel." So shall the bodies of the children of the resurrection shine in the other world, when "the bodies of all saints are made like unto Christ's glorious body."[53]

Thirdly, our bodies shall be raised in power. Presently, our mortal bodies are "no better than clogs and fetters," binding and confining our souls. As St. Augustine frequently insisted, "the corruptible body presses down the soul, and the earthly tabernacle weighs down the mind."[54] In the resurrection, however, our bodies, having been changed to immortal and incorruptible bodies, also shall be made sprightly and nimble. As such, they shall "be obedient and able instruments of the soul." Then, they shall "mount up with wings as eagles; they shall run, and not be weary; they shall walk, and not faint."[55] In fact, the saints shall meet the Lord in the air when he comes to judgment, and mount up with him into the highest heaven.[56] The mortal body presently is "slow and heavy in its motions, listless and soon tired with action. But our heavenly bodies shall be as fire; as active and nimble as our thoughts are."[57]

Fourthly, our resurrected bodies will be "spiritual bodies." In the mortal condition, our spirits are forced to serve our bodies. But in the other world, our resurrected bodies "shall then wholly serve our spirits, and minister to them," Wesley reasoned.[58] Ironically, after a mortal existence in which the spirit was dependent upon the body, the body shall be dependent upon the spirit in the immortal life. Wesley's statement of contrast on this point is insightful -

"By a 'natural body' we understand one fitted for this lower, sensible world, for this earthly state; so 'a spiritual body' is one that is suited to a spiritual state, to an invisible world, to the life of angels. And, indeed, this is the principal difference between a mortal and a glorified body."[59]

Before leaving this contrast between mortal and glorified bodies, we should observe that Wesley introduced a definitive statement concerning "the flesh." Being too much an Augustinian in theology, Wesley rejected a Manichaean explanation of "flesh." Following Augustine, he argued that "the flesh" is our experience of mortality in which the body presses down the soul, where the spirit is in subjection to the body and its desires. This is "the flesh" of which St. Paul spoke, and against which he wrestled. It is the "most dangerous enemy we have: We therefore deny and renounce it in our baptism."[60] Its dangers are numerous -

"It constantly tempts us to evil. Every sense is a snare to us. All its lusts and appetites are insubordinate. It is ungovernable, and often rebels against reason. The law in our members wars against the law of our mind. When the spirit is willing, the flesh is weak; so that the best of men are forced to keep it under, and use it hardly, lest it should betray them into folly and misery. And how does it hinder us in all our devotions! How soon does it jade our minds when employed on holy things! How easy, by its enchanting pleasures, does it divert them from those noble exercises! But when we have obtained the resurrection unto life, our bodies will be spiritualized, purified, and refined from their earthly grossness; then they will be fit instruments for the soul in all its divine and heavenly employment; we shall not be weary of singing praises to God through infinite ages."[61]

(3) Wesley's inferences from this whole presentation on resurrection. First, Wesley inferred, pilgrims must prepare themselves for living in glorified bodies - "by cleansing ourselves more and more from all earthly affections, and weaning ourselves from this body, and all the pleasures that are peculiar to it."[62] Secondly, pilgrims must realize that there are degrees of glory in eternity - "For although all the children of God shall have glorious bodies, yet the glory of them all shall not be equal. 'As one star differeth from another star in glory, so also is the resurrection of the dead.'" Wesley's words seem enigmatic to contemporary clerics who try to read democratic principles into every nook and cranny of theology. What kind of other world can eternity be when the children of God are not equal in glory? How can a God of love allow such crass discriminations? Wesley stills the critics by applying scriptural principles rather than political/ideological ones -

"They shall all shine as stars; but those who, by a constant diligence in well-doing, have attained to a higher measure of purity than others, shall shine more bright than others. They shall appear as more glorious stars. It is certain that the most heavenly bodies will be given to the most heavenly souls; so that this is no little encouragement to us to make the greatest progress we possibly can in the knowledge and love of God, since the more we are weaned from the things of the earth now, the more glorious will our bodies be at the resurrection."[63]

Thirdly, Wesley exhorted, we pilgrims are "on our journey towards home, and we must expect to struggle with many difficulties;
but it will not be long ere we come to our journey's end, and that
will make amends for all."[64] Indeed, he believed, when Christ shall
descend from heaven with a shout, when the archangel and the
trumpet of God shall sound above all other voices and affirmations,
when lightning flashes shall split the heavens as shafts of great
light, when the clouds shall form about the all-powerful Son of
man, suddenly, in that teleological moment, we shall all be changed
- this mortal shall put on immortality, and this corruptible shall put
on incorruption! Then the resurrection long promised shall be an
accomplished fact!

THE MILLENNIUM AND JUDGMENT

Wesley's understanding of the themes of the millennium and the
judgment should be seen against a long and popular tradition of
prophetic fascination. Christian chiliasm, after the time of Emperor Constantine, died rapidly as a vital concern of Christians.
However, during the Middle Ages, it revived in the West, becoming
popular in every segment of society. One generation passed its
millenarian hope on to the next, identifying social ills as the work
of an all-destroying demon - as robbery, rapine, torture and massacre. But these things were also "signs" of the approaching Second Coming and the Kingdom of the Saints. Just before these
come, however, there was to be a "final time of troubles" or "a great
tribulation" such as the world had never known before, featuring
bad rulers, wars, civil disorders, drought, famine, plague, comets,
sudden deaths of prominent persons and an increase in general sin-
fulness. Invading armies of Huns, Magyars, Mongols, Saracens and
Turks were always "signs" of the nearness of the end. These armies
were frequently called "hordes of antichrists" and "Gog and Magog." Then too, any ruler who became a tyrant was bound to be
likened to Antichrist, the son of perdition. Such a ruler was usually
described in royal chronicles as *rex iniquus*. Upon the death of a
tyrant ruler, indicating that he had not fulfilled prophecy, he was

degraded to the lesser rank of "precursor," and the waiting for the real *rex iniquus* to be revealed would begin all over again.[65] To digress momentarily, twentieth century chiliasm, in American Protestantism, is guilty of the same branding of rulers as the Antichrist, then on their passing, quickly naming a replacement - Kaiser Wilhelm, Hitler, Mussolini, Stalin, Khrushchev, and even Henry Kissinger who is still living in 1989.

When the eighteenth century arrived, the popular and superstitious millennial tradition of the Middle Ages was still strong, especially in England. Few of its supporters knew the biblical tradition that spawned the doctrine, although they claimed they did. Wesley found their explanations to be half-truths at best, partly taken out of context from the Bible and often mixed with extraneous and non-biblical material. On the contemporary scene, a similar situation may be found.

Consequently, Wesley wished to rescue his Methodist pilgrims from such sleazy chiliasm. His EXPLANATORY NOTES UPON THE NEW TESTAMENT takes great pains to clarify difficult eschatological concepts, especially that of the *millennium*. The term millennium comes from the Latin *mille*, meaning "a thousand." The Greek equivalent is *chilia*, from which comes the term "chiliasm." The concept of a special age of one thousand years is found only in the Apocalypse, chapter XX, 1-10. However, the passage is hardly simple and uncomplicated, as some interpreters attempt to make it appear. Wesley's analysis of it goes deep, extracting from the text only what is there and then drawing rational conclusions from what the text itself says. Let popular tradition be damned when it misses the meaning of Scripture.

The Millennial Passage of Revelation XX, 1-10. Wesley's commentary runs verse by verse, treating the more important words and phrases. These are the crucial ones -

Verse 1 - "*And I saw an angel descending* - Coming down with a commission from God. Jesus Christ Himself overthrew the beast: the proud dragon shall be bound by an angel; even as he and his angels were cast out of heaven by Michael and his angels. *Having the key of the bottomless pit* - Mentioned before, Rev. IX, 1. *And a great chain in his hand - The angel of the bottomless pit* was shut up therein before the beginning of the first woe. But it is now first Satan, after he had occasioned the third woe, is both chained and shut up."

Verse 2 - "*And he laid hold on the dragon* - With whom undoubtedly his angels were now cast into the bottomless pit, as well as finally 'into the everlasting fire' (Matt. XXV, 41). *And bound him a thousand years* - That these thousand do not precede, or run parallel with, but wholly follow, the times of the beast, may manifestly appear, (1) From the series of the whole book, representing one continued chain of events. (2) From the circumstances which precede. The woman's bringing forth is followed by the casting of the dragon out of heaven to the earth. With this is connected the third woe, whereby the dragon through, and with, the beast, rages horribly. At the conclusion of the third woe the beast is overthrown and cast into 'the lake of fire.' At the same time the other grand enemy, the dragon, shall be bound and shut up. (3) These thousand years bring a new, full, and lasting immunity from all outward and inward evils, the authors of which are now removed, and an affluence of all blessings. But such a time the Church has never yet seen. Therefore it is still to come. (4) These thousand years are followed by the last times of the world, the letting loose of Satan, who gathers together Gog and Magog, and is thrown to the beast and false prophet 'in the lake of fire." Now Satan's accusing the saints in heaven, his rage on earth, his imprisonment in the abyss, his seducing Gog and Magog, and being cast into the lake of fire, evidently succeed each other. (5) What occurs from Rev. XX, 11 to XXII, 5 manifestly follows the things related in the nineteenth chapter. The thousand years came between; whereas if they were past, neither the beginning nor the end of them would fall within this period. In a short time those who assert that they are now at hand will appear to have spoken the truth. Meantime, let every man consider what kind of happiness he expects therein. The danger does not lie in maintaining that the thousand years are yet to come; but in interpreting them, whether past or to come, in a gross and carnal sense. The doctrine of the Son of God is a mystery. So is his cross; and so is His glory. In all these He is a sign that is spoken against. Happy they who believe and confess Him in all!"[66]

Verse 3 - "*And set a seal upon him* - How far these expressions are to be taken literally, how far figuratively only, who can tell? *That he might deceive the nations no more* - One benefit only here is expressed, as resulting from the confinement of Satan. But how many and great blessings are implied: For the grand enemy being removed, the kingdom of God holds on its uninterrupted course among the nations; and the great mystery of God, so long foretold, is at length fulfilled; namely, when the beast is destroyed, and Satan bound. This fulfillment approaches nearer and nearer; and contains things of the utmost importance, the knowledge of which becomes every day more distinct and easy. In the meantime it is highly necessary to guard against the present rage and subtilty of the devil. Quickly he will be bound: when he is loosed again, the martyrs will live and reign with Christ. Then follow His coming in glory, the new heaven, new earth, and new Jerusalem."[67]

Before departing from our exegetical survey, Wesley's statement - "the martyrs will live and reign with Christ. Then follow His coming in glory" - needs some attention. Because of this position, he has sometimes been called a "Post-millennialist." Actually, he was an Augustinian with the soul being in the first resurrection. In treating verse 4, he argued that "two distinct thousand years are mentioned throughout this whole passage." Each is mentioned three times - (1) the thousand wherein Satan is bound (verses 2, 3, 7), and (2) the thousand wherein the saints shall reign (verses 4a, 4b, 6). "The former end before the end of the world; the latter reach to the general resurrection. So that the beginning and end of the former thousand is before the beginning and end of the latter." Wesley added, "During the former, the promises concerning the flourishing state of the Church (Rev. X, 7) shall be fulfilled." As for the latter, "While the saints reign with Christ in heaven, men on earth will be careless and secure."[68] The martyred saints "reign with Christ in heaven" for "the" thousand years. Meanwhile, "men on earth" are safe and free from cares. Moreover, Wesley argued, the general resurrection comes at the end of the thousand years of the heavenly reign, the thousand years of Satan's binding having begun and ended "much sooner." The "second thousand years, begin at the same point immediately after the first thousand." Then Wesley added a very important interpretation - "But neither the beginning of the first nor of the second thousand will be known to the men upon earth, as both the imprisonment of Satan and his loosing are transacted in the invisible world."[69] Hence the millennial reign of the martyrs with Christ in heaven takes place in the invisible world as well, not being known by men on earth. Wesley believed that this interpretation was sound -

> "By observing these two distinct thousand years many difficulties are avoided. There is room enough for the fulfilling of all the prophecies, and those which before seemed to clash are reconciled; particularly those which speak, on the one hand, of a most flourishing state of the Church as yet to come; and, on the other, of the fatal security of men in the last days of the world."[70]

At the close of Wesley's commentary on the Apocalypse, he added a time chart, indicating the various events of the book in reference to the Christian calendar. His account of what will occur prophetically in A.D. 1836 helps clarify this puzzle of the millennium -

> "1836 - The end of the non-chronos, and of the many kings; the fulfilling of the word, and of the mystery of God; the repentance of the survivors in the great city; the end of the 'little time,' and of the three times and a half; the destruction of the beast; the imprisonment of Satan. Afterward: The loosing of Satan for a small time; the beginning of the thousand years' reign of the saints; the end of the small time. The end of the world; all things new."[71]

Let us assume that Wesley really believed this interpretation of the millennium being heavenly rather than an earthly phenomenon. In such a case, his unusually optimistic sermon, THE GENERAL SPREAD OF THE GOSPEL, has a much greater evangelical significance. If Christ and the martyred saints actually ruled on earth, even with Satan being bound, conversions would be somewhat forced conversions, else Christ would not be much of a ruler. But, if the millennial reign is in heaven, with the peoples of the earth no longer under the dragon's influence, and Christians still live on earth during that time, until the next resurrection, Wesley's sermon point is a good one - this bit of time will be "a good Pentecost" as he termed it, in which the Church will become effective in preaching and living the gospel. Then mass conversions will be the rule, and Jews, Mahometans, heathens and nominal Christians will "become as little children" by happy conversions. The sermon will be studied shortly.

While the point has been made several times that Wesley followed Bengelius in exegeting Revelation, it should also be noted that millennial interpretations were numerous and some very ancient. Justin Martyr (A.D. 100-165) held a distinctive view of that age, which Wesley wrote about in his famous letter to the Rev. Dr. Middleton -

> "The doctrine (as you very well know) which Justin deduced from the Prophets and the Apostles, and in which he was undoubtedly followed by the Fathers of the second and third centuries, is this: - The souls of them who have been martyred for the witness of Jesus, and for the word of God, and who have not worshipped the beast, neither received his mark, shall live and reign with Christ a thousand years. But the rest of the dead shall not live again, until the thousand years are finished. Now to say they believed this, is neither more nor less than to say, they believed the Bible." [72]

However, there is a monumental problem with Wesley's citing of indebtedness to Justin Martyr. In his DIALOGUE WITH TRYPHO, Justin did indeed describe the Christian doctrine of the millennium, but he had the reign of Christ as an earthly one, from Jerusalem. [73] Luke Tyerman called attention to this difference over a century ago. [74]

The challenge of trying to move through another man's mind, determining what he thought at a given moment of time, with his writings as your window, is often frustrating and usually impossible. This aspect of Wesley's theology represents that kind of a challenge. It seems that Wesley shifted ground on numerous theological positions, always with one foot on the ground and the other in the air. It may well be that the millennium was one of these subjects. When he completed the *EXPLANATORY NOTES UPON THE NEW TESTAMENT*, in 1754, he was convinced that Bengelius was correct on the millennium being a heavenly rule, and he advanced that position in his commentary on Revelation. But, as time went on, he sought the works of other writers on the subject, probably feeling that his treatment in the *NOTES* might need revision. In 1762, he read the treatise of Charles Perronet on II Peter III, 13 (". . .we look for new heavens and a new earth"). One particular tenet of the treatise was the idea that the millennial reign of Christ was to be a heavenly affair and not earthly.[75] No doubt Wesley was relieved concerning the position he had posited in the *EXPLANATORY NOTES*. Two years later, he corresponded with the Rev. Dr. Thomas Hartley, author of a new book entitled, *PARADISE RESTORED: OR A TESTIMONY TO THE DOCTRINE OF THE BLESSED MILLENNIUM*. The treatment given the subject by Hartley stimulated Wesley greatly. He wrote - "I cannot but thank you for your strong and seasonable confirmation of that comfortable doctrine; of which I cannot entertain the least doubt as long as I believe the Bible."[76] It was from Hartley's book that Wesley's views on the millennium began to change, and the results were startling. In 1784, Wesley placed a series of articles in his *ARMINIAN MAGAZINE*, turning the millennial reign in heaven into a "middle state" on earth. He now referred to the millennium as a -

> "middle state betwixt the present pollution, corruption, and degradation of this terrestrial mansion; and that of a total, universal restoration of all things, in a purely angelical, celestial, ethereal state. Now this middle state, betwixt these two extremes, can be no other, than the paradisiacal state of the earth renewed and restored to its primitive lustre and beauty."[77]

Wesley's new millennium, as a middle state between gross imperfection and ultimate perfection, is a period of restoration to primitive goodness and beauty. It is a time when nations convert swords and spears into plowshares and pruninghooks, and they war no more. It is a time when the wolf and lamb feed together and the lion eats straw like the ox.[78] From beginning to end, the middle state shall be a time of "universal, permanent felicity" - as long as it lives, it will be marked by blessed and expanding peace.[79] By 1784, then, Wesley placed his "foot in the air" on new ground, and, in doing so, he lifted his other foot from where once it was planted. Although he shifted in view, he believed that the biblical evidence for this kind of millennium was overwhelming, better revealing the divine plan of bringing the fallen creation to a restoration before it is made perfect and eternal.

Whether Wesley's earlier millennium or his later millennium is the proper interpretation of the reign of Christ, either in heaven or on earth, his sermon on *THE GENERAL SPREAD OF THE GOSPEL* is an indication of what he believed would take place on earth prior to the beginning of the great judgment. In the standard collection, this sermon is numbered LXIII. Its chief text is from Isaiah XI, 9 - "The earth shall be full of the knowledge of the Lord, as the waters cover the sea." Beginning with a pertinent observation - "In what a condition is the world at present!" - Wesley the preacher advanced through a statistical treatment of the peoples of the earth and the chief religions among them. He believed that the Christian faith is superior to all the other faiths because it has "abundantly more knowledge... more scriptural and more rational modes of worship."[80] Yet the Christian religion is divided within itself, and within the west the Papists and the Protestants generally are both far from being true Christians. "Such is the present state of mankind in all parts of the world!" Wesley concluded. He then marveled -

"But how astonishing is this, if there is a God in   heaven, and if his eyes are over all the earth! Can despise the work of his own hand? Surely this is one of the greatest mysteries under heaven! How is it possible to reconcile this with either the wisdom or goodness of God? And what can give ease to a thoughtful mind under so melancholy a prospect? What but the considera- tion, that things will not always be so; that another scene will soon be opened? God will be jealous of his honour: He will arise and maintain his own cause. He will judge the prince of this world, and spoil him of his usurped dominion. He will give his Son 'the Heathens for his inheritance, and the uttermost parts of the earth for his possession.' The earth shall be filled with the knowledge of God, producing uniform, uninterrupted holiness and happiness, shall cover the earth; shall fill every soul of man."[81]

Such a universalism, however, is not completely universal. Wesley was careful to maintain that this future state brings the knowledge of God only to those who live during that time, and even then, their faith response to divine grace is required as the appropriate response. God does not change his way of receiving sinners -

> "As God is One, so the work of God is uniform in all ages. May we not then conceive how he will work on the souls of men in the times to come, by considering how he does work now, and how he has wrought in times past?"[82]

The best commentary on God's "general manner of working" salvation, Wesley thought, is the history of Methodism. Wesley gave a Latin motto to his movement - *Qui fecit nos sine nobis, non salvabit nos sine nobis* - "He that made us without ourselves, will not save us without ourselves."[83] An historical description of the rise of Methodism, with God's methodology of salvation, is considered next in the sermon. The spread of the evangelical revival to many parts of Christendom led Wesley to express his great expectation that shortly even the Church of Rome would be included -

> "May we not suppose that the same leaven of pure and undefiled religion, of the experimental knowledge and love of God, of inward and outward holiness, will afterwards spread to the Roman Catholics in Great Britain, Ireland, Holland: in Germany, France, Switzerland: and in all other countries where Romanists and Protestants live intermixed and familiarly converse with each other?"[84]

Perhaps the hardest persons to convert, Wesley reasoned, are the "wise and learned," "men of genius" or "philosophers." In the special time coming, even these shall be converted and "become as little children." With them, of course, will be the rich, rulers, princes, and kings of the earth.[85] In that time, no one shall say, "Know the Lord!" Quoting from the prophet Jeremiah (XXXI, 34), Wesley gave the reason - "For they shall all know me, from the least to the greatest."

This special time in the near future was called "the grand Pentecost" by Wesley. When it is "fully come... devout men in every nation under heaven, however distant in place from each other, shall all be filled with the Holy Ghost."[86] These righteous persons will "continue steadfast in the Apostles' doctrine, and in the fellowship, and in the breaking of bread, and in prayers." Moreover, Wesley said, "the natural, necessary consequence of this will be that the same as it was in the beginning of the Christian Church" - even to the holding of goods and possessions in common.[87] The Christian Church then will be truly of one mind and of one accord, without divisions. It will be compassionate, loving and sharing sacrificially.

Wesley's sermon on the spread of the gospel claims that when such a restoration of primitive Christianity occurs, then the traditional stumbling-blocks of centuries will cease to exist. Then the Mahometans will be converted -

> "The Mahometans will look upon them (Christians) with other eyes, and begin to give attention to their words. And as their words will be clothed with divine energy, attended with the demonstration of the Spirit and of power, those of them that fear God will soon take knowledge of the Spirit whereby the Christians speak. They will 'receive with meekness the engrafted word,' and will bring forth fruit with patience. From them the leaven will soon spread to those who, till then, had no fear of God before their eyes. Observing the *Christian dogs*, as they used to term them, to have changed their nature; to be sober, temperate, just, benevolent; and that, in spite of all provocations to the contrary, from admiring their lives, they will surely be led to consider and embrace their doctrine."[88]

In similar fashion, Wesley said, the stumbling-block so long laid before the heathens will be removed. In that pentecostal time, no more will a Malabrian heathen say - "Christian man take my wife: Christian man much drunk: Christian man kill man! *Devil Christian*! me no Christian."[89] As for the heathens living so remotely from Christians, without any knowledge whatsoever of the faith of Jesus Christ, God shall use many ways of bringing them to salvation by the preaching of the gospel.[90] Then shall the promise be fulfilled - "He will give his Son 'the uttermost parts of the earth for his possession."[91]

What of Israel? Wesley included the Jews in this universal turning to God. His words on the subject are drawn from many biblical sources -

"And so all Israel too shall be saved. For 'blindness has happened to Israel,' as the great Apostle observes, (Rom. XI, 25, etc.) till the fullness of the 'Gentiles be come in.' Then 'the Deliverer that cometh out of Sion shall turn away iniquity from Jacob.' 'God hath now concluded them all in unbelief, that he may have mercy upon all.' Yea, and he will so have mercy upon all Israel, as to give them all temporal with all spiritual blessings. For this is the promise: 'For the Lord thy God will gather thee from all nations, whither the Lord thy God hath scattered thee. And the Lord thy God will bring thee into the land which thy fathers possessed, and thou shalt possess it. And the Lord thy God will circumcise thine heart, and the heart of thy seed, to love the Lord thy God with all thine heart, and with all thy soul." (Deut. XXX, 3, etc.)"[92]

Of all the passages written by John Wesley concerning the Church, none is as powerful and significant as the one closing this sermon. Usually the term "Church Triumphant" is related only to the Church already beyond time. However, Wesley changed that condition, making the Church, in this special segment of time before the great judgment, the "Church Triumphant." The passage reads -

"At that time will be accomplished all those glorious promises made to the Christian Church, which will not then be confined to this or that nation, but will include all the inhabitants of the earth. 'They shall not hurt nor destroy in all my holy mountain.' (Isa. XI, 9) 'Violence shall no more be heard in thy land, wasting nor destruction within thy borders; but thou shalt call thy walls salvation, and thy gates Praise.' Thou shalt be encompassed on every side with salvation, and all that go through thy gates shall praise God. 'The sun shall be no more thy light by day; neither for brightness shall the moon give light unto thee: But the Lord shall be unto thee an everlasting light, and thy God thy glory.' The light of the sun and moon shall be swallowed up in the light of His countenance, shining upon thee. 'Thy people also shall be all righteous. . . the work of my hands, that I may be glorified.' 'As the earth bringeth forth her bud, and the garden causeth the things that are sown in it to spring forth; so the Lord God will cause righteousness and praise to spring forth before all the nations.' (Isa. LX, 18, etc. and Isa. LXI, 11.)"[93]

The concluding remarks of the preacher of this sermon pull together all the eschatological themes important to the Christian faith of early Methodism -

"All unprejudiced persons may see with their eyes, that He is already renewing the face of the earth: And we have strong reason to hope that the work he hath begun, he will carry on unto the day of the Lord Jesus; that he will never intermit this blessed work of his Spirit, until he has fulfilled all his promises, until he hath put a period to sin, and misery, and infirmity, and death, and reestablished universal holiness and happiness, and caused all the inhabitants of the earth to sing together, 'Hallelujah, the Lord God omnipotent reigneth!' 'Blessing, and glory, and wisdom, and honour, and power, and might, be unto God for ever and ever!' (Rev. VII, 12)."[94]

The mature understanding of Wesley of this special time, that he called the "middle state," or the "millennium," places it in time prior to the coming of Christ in glory and judgment. His sermon makes this clear. For this reason some students of Wesley refer to his position as being post-millennial. However, when one follows the sequence of events in his sermon *THE GREAT ASSIZE*, a radically different scheme emerges. This is the case because the latter sermon on the great judgment was written and preached long before 1784 and his shifting of millennial views. Nevertheless, the *GREAT ASSIZE* sermon is valuable for gaining an appreciation of what Wesley believed would take place in the final judgment, regardless of when it might occur.

When one thinks of Wesley's preaching and writing ministries, the subject of divine judgment always surfaces. The majority of his sermons inject the theme, and his wide-ranging writings abound with references to God giving justice and mercy in one way or another. But the most concentrated treatment on the subject is that sermon, *THE GREAT ASSIZE*, preached by Wesley at Bedford, on Friday, March 10, 1758.[95] An analysis of this sermon is indispensable, although some references no longer fit the later views of Mr. Wesley. However, that is not the case in reference to the theme of judgment itself.

The Great Assize Sermon was preached in a packed courtroom. The judge sat up high above the courtroom floor, and he was dressed in judicial regalia as an emblem of royal justice. The lawyers, both prosecutors and advocates, sat at lesser desks, ready to accuse and defend. The defendants were all there, under guard, awaiting their turn at the bar of justice. Witnesses, many of them, crowded the room, ready to tell the truth, the whole truth, and nothing but the truth, some for and some against the defendants. Bailiffs also were present to keep order and administer the lesser mechanics of the court. Spectators were there, some with deep interest in various cases, and others there simply out of curiosity. Then, the Rev. Mr. John Wesley, Priest of the Church of England and late Fellow of Lincoln College, Oxford University, was introduced to the court, for he was to preach a sermon to launch the new court session beginning that day.

Wesley used the court scene about him as a picture of a more awesome court to come - "For yet a little while, and 'we shall all stand before the judgment seat of Christ.'" Furthermore, in that day, "every one of us shall give an account of himself to God." His sermon developed three main points: (1) the chief circumstances which will precede our standing before the judgment seat of Christ; (2) The judgment itself; and (3) A few of the circumstances which will follow it. A survey of these points - being a synthesis of the gospels, epistles, and Apocalypse - will prove helpful.

(1) The chief circumstances which will precede our standing before the judgment seat of Christ. These "circumstances" are "signs" in earth and heaven, some of which were treated earlier in this chapter - earthquakes in all places, thunders and lightnings, sun and moon and stars not giving light, and stars falling from heaven. Some signs, mentioned here by Wesley, would not be included in a sermon revised along the line of his late millennial views - "every island shall flee away, and the mountains will not be found," and "the air will be all storm and tempest, full of dark vapours and pillars of smoke."[96] Then, after the signs, the coming of Christ occurs, as described by St. Paul in I Thessalonians IV - "Then shall be heard the universal shout, from all the companies of heaven, fol-

lowed by the 'voice of the archangel,' proclaiming the approach of
the Son of God and Man." The trumpet of God shall then sound,
alarming all that sleep in the dust of the earth. The young Wesley
quickly blended the Apocalypse with Paul - "All the graves shall
open, and the bodies of men arise. The sea also give up the dead
which are therein, and everyone shall rise with 'his own body.'"[97]
Wesley reveals his philosophical commitment to Aristotelian metaphysics at this point when he affirmed that "his own body" means
"his own in substance, although so changed in its properties as we
cannot now conceive."[98] The sermon continues - "So that all who
ever lived and died, since God created man, shall be raised incorruptible and immortal." At this very time, the Son of man shall
send forth his angels "over all the earth; and they shall gather his
elect from the four winds, from one end of heaven to the other"
(Matt. XXIV, 31). And seated upon his throne of glory, with all of
the nations gathered before him, "he shall separate them one from
another, and shall set the sheep, 'the good,' on his right hand, and
the goats, 'the wicked,' upon the left" (Matt. XXV, 31). Quoting
from St. John, Wesley added, "And the books were opened. . . and
the dead were judged out of those things which were written in the
books."[99]

(2) The judgment itself - The person appointed by God to be the
judge at this great trial is Jesus Christ. His "day" is the day of judgment, but with him "a day is as a thousand years." The ancient Fathers of the Church, Wesley claimed, believed that this day of
judgment, conducted by Christ, lasts for a thousand years. It is interesting to note that the millennial idea can apply to some activity
other than reigning or restoring. Wesley, in 1758, was content to
equate the millennium with the judgment day. He was personally
committed to this judgment spanning a thousand years or two -

"The ancient Fathers drew that inference, that, what is commonly called the
day of judgment would be indeed a thousand years: And it seems they did
not go beyond the truth; nay, probably they did not come up to it. For, if we
consider the number of persons who are to be judged, and of actions which
are to be inquired into, it does not appear, that a thousand years will suffice

for the transactions of that day; so that it may not improbably comprise several thousand years."[100]

Where will the judgment take place? Wesley acknowledged that many think it will be held on earth, but he was not convinced of that. Wesley used St. Augustine's exegesis of I Thessalonians IV, 17. Augustine wrote -

"We shall be caught up together with them in the clouds, to meet the Lord in the air, refers to the living Christians and the resurrected Christians meeting Christ in the air, above the earth, and there helping Christ in the judging of the rest of humanity on earth - 'He shall catch them up into the air' meaning, 'He shall lift up to seats of judgment.'"[101]

Wesley's version of Augustine's argument is as follows -

"St. Paul writes to the Thessalonians: 'The dead in Christ shall rise first. Then we who remain alive, shall be caught up together with them, in the clouds, to meet the Lord in the air.' . . .So that it seems most probable, the great white throne will be high exalted above the earth."[102]

The persons to be judged are beyond numbering, since they constitute all humans that have ever lived. Everyone shall hear the voice of the great judge - "All universally, all without exception, all of every age, sex, or degree; all that ever lived and died, or underwent such a change as will be equivalent with death."[103] In the judgment before Christ, Wesley noted, human greatness means little, because just before that day "the phantom of human greatness disappears, and sinks into nothing."[104] Every person will then, truthfully and honestly, give an account of all "his own works" done in his body, whether good or evil. But not only deeds will be judged, words will too. Every word ever spoken, with intent and consequence, will be tried by the judge who once was "about our bed, and about our path, and spieth out all our ways!"[105] The very depth of our souls will be examined by the judge - "every appetite, passion, inclination, affection, with the various combinations of them, with every temper and disposition that constitute the whole

complex character of each individual."[106] By so close an examination it shall "be clearly and infallibly seen, who was righteous, and
who was unrighteous; and in what degree every action, or person,
or character was either good or evil."[107]

Then the judge, who is Christ the King, shall declare those at his
right to be worthy of his kingdom. He shall recite their deeds and
words before men and angels, although they had been "unknown or
forgotten among men." All their former sufferings for Christ's sake
shall be revealed by the judge. He shall bestow "honour before
saints and angels" upon each of them.[108] But, Wesley asked, what
of the sins of the righteous? Will these be remembered in that
day? "Many believe they will not. . . seeing they would still have
sorrow, and shame, and confusion of face to endure." These advocates further argue that the recitation of the sins of the righteous at
the judgment would be inconsistent with God's promise through
Jeremiah - "I will forgive their iniquities, and remember their sin no
more" (Jer. XXXI, 34). Wesley's response to this problem of sin in
saints is remarkable, if not surprising -

> "It may be answered, It is apparent and absolutely necessary, for the full dis-
> play of the glory of God; for the clear and perfect manifestation of his wis-
> dom, justice, power and mercy, toward the heirs of salvation; that all the cir-
> cumstances of their life should be placed in open view, together with their
> tempers, and all the desires, thoughts and intents of their hearts: Otherwise,
> how would it appear out of what a depth of sin and misery the grace of God
> had delivered them? And, indeed, if the whole lives of all the children of men
> were not manifestly discovered, the whole amazing contexture of divine
> providence could not be manifested; nor should we yet be able, in a thousand
> instances, 'to justify the ways of God to man.' Unless our Lord's words were
> fulfilled in their utmost sense, without any restriction or limitation - 'There is
> nothing covered that shall not be revealed, or hid that shall not be known"
> (Matt. X, 26).[109]

As the righteous, therefore, stand in the judgment, hearing their
sins as well as their goodness, they shall be judged with the maximum of justice, mercy, and truth - "All the transgressions which
they had committed shall not be once mentioned unto them to

their disadvantage; that their sins, and transgressions, and iniquities shall be remembered no more to their condemnation."[110] The children of God will find this treatment comforting, "far from feeling any painful sorrow or shame," and they "will rejoice with joy unspeakable."[111]

When all the elect or righteous have been judged, Wesley said, then the remainder of humanity will be judged, standing to the left of the judge-king. Their accounting will be as thorough as that of the righteous before them. Outward and inward deeds, of body, soul, and spirit will be examined in the greatest detail. When Christ has finished with them, he shall give "the joyful sentence of acquittal... upon those upon the right hand; (and) the dreadful sentence of condemnation upon those on the left; both of which must remain fixed and unmovable as the throne of God."[112]

(3) A few of the circumstances which will follow the general judgment - First, the sentences are executed. The evil "shall go away into eternal punishment." On the other hand, "the righteous (go) into life eternal." Flailing at those who say the punishment is only temporary, Wesley added - "It follows, that either the punishment lasts for ever, or the reward too will come to an end: - No, never, unless God could come to an end, or his mercy and truth could fail."[113] The reward of the wicked has several parts, according to Wesley. They are "turned into hell" - hell now being synonymous with "the lake of Fire" - far removed from "the presence of the Lord, and from the glory of his power." There they are subjected to burning "brimstone," originally prepared for the devil and his angels. Consequently, they "gnaw their tongues for anguish and pain, they will curse God and look upward." There "the dogs of hell - pride, malice, revenge, rage, horror, despair - continually devour them." Moreover, they "have no rest, day or night, but the smoke of their torment ascendeth for ever and ever!" "Their worm dieth not, and the fire is not quenched."[114] Secondly, the present creation will then pass away.[115] And thirdly, a new creation will appear - "So great shall the glory of the latter (creation) be!"[116]

With rational appeals for his audience at court to prepare for the great judgment, Wesley closed his sermon.

As important as the sermon *THE GREAT ASSIZE* is for Wesleyan theology, it is actually but an outline of the subject. Our task requires a deeper inquiry than that of the sermon.

The eighteenth century in England was an age of religious controversies. In reference to the great judgment, heaven and hell, there were extreme positions, and many shades of interpretation between the poles. Some "enlightened men" rejected the whole scheme as a carry-over of primitive superstition. A new breed of Calvinists, however, took the judgment, heaven and hell, seriously, contending that Christ's atonement was particular, that is, for the "elect" alone. Hence, at the great judgment the elect are automatically justified and rewarded with eternal life. They are the elect, because, before the foundations of the world, God predestined them to be saved, and consequently, God also predestined the non-elect to be damned. Wesley fought this view as being pernicious and false. If it were true, the great judgment is actually meaningless, and every non-Calvinist knows it. One popular predestinarian argument, following St. Paul's analogy of the potter and the clay, reasoned that the creator has the right to make his creatures as he wishes and to dispose of them the same way. Wesley found such a misunderstood analogy absurd. So, in 1782, Wesley attacked this argument in his *ARMINIAN MAGAZINE*, quoting a poem by Dr. W. Byrom -

> "Hath not the potter power to make his clay
> Just what he pleases? - Well. And tell me pray
> What kind of potter must we think a man,
> Who does not make the best of it he can?
> Who, making some fine vessels of his clay,
> To shew his power, throws all the rest away,
> Which, in itself, was equally fine?
> What an idea this of power divine!"[117]

Wesley, with his commitment to general atonement - that Christ died for all mankind, not willing that any should perish - saw the great judgment in terms of divine grace extended through Christ, as "preventing grace," to all sons and daughters of Adam. Such an ex-

tension, however, does not make a Christian automatically. Hardly.
Salvation for Wesley was not an event but rather a process, with
stages and degrees.[118] Even preventing grace implies "some degree of salvation,"[119] but it is only preparatory for higher works of
grace. Where sin abounds, grace does much more abound, Wesley
believed. Human response to all forms of grace is faith, and faith
too comes in varying degrees and stages - preliminary faith, justifying faith, and sanctifying faith.[120] Therefore, the great judgment
for Wesley meant something radically different than what it meant
for predestinarians of his day. Hence Wesley wrote of the fairness
of God with every man in the great judgment -

> "The general rule stands firm as the pillars of heaven: 'The judge of all the
> earth will do right. He will judge the world in righteousness,' and every man
> therein, according to the strictest justice. He will punish no man for doing
> anything which he could not possibly do. Every punishment supposes the of-
> fender might have avoided the offence for which he is punished: Otherwise,
> to punish him would be palpably unjust, and inconsistent with the character
> of God our Governor."[121]

When the great judgment begins, Wesley believed, the order of
cases the judge will hear begins with the righteous of all ages,
starting with those of antiquity and chronologically coming down
into more recent times. No Methodist will appear immediately, but
only very late in the proceedings. God revealed the last judgment
to his people in the time of Enoch (Genesis V, 18ff).[122] Since then
the righteous of every generation have been preparing for it, except some have not taken its inevitability seriously, living outside
the state of holiness. The Church of apostolic times had members
who scoffed at the idea of Christ's coming with judgment.[123] St.
Peter described them as having fallen from the true liberty into
spiritual slavery -

> "For if after they have escaped the pollutions of the world through the
> knowledge of the Lord and Saviour Jesus Christ, they are again entangled
> therein, and overcome, their last state is worse than the first. For it had been
> better for them not to have known the way of righteousness, than, having

known it, to turn from the holy commandments delivered to them. But it has befallen them according to the true proverb, The dog it turned to his own vomit, and the sow that was washed to her wallowing in the mire." (II Peter II, 20-22).

Wesley often preached on this theme, citing many biblical examples of persons who professed faith in Christ but lacked the holiness of heart and life. He was bold to refer to this lifestyle as "backsliding." Preaching against such a popular pastime was not original with Wesley. The Church of England, since the time of Martin Bucer's revision of the First Prayer Book of King Edward VI (A.D. 1549), condemned the practice. When the Church of England established the HOMILIES, shortly after the Prayer Book's revision, Homily VIII treated the phenomenon under the title of *DECLINING FROM GOD*. Some of the classic statements from the *HOMILY* are worthy of attention -

"Pride is the fountain of all sin. And as by pride and sin we go from God, so shall God, and all goodness with him, go from us. As touching our turning... from God, you shall understand that it may be done divers ways. Sometimes directly by idolatry. . . sometimes men go from God by lack of faith and mistrusting of God."[124]

"Some by the neglecting of their neighbours; some by not hearing of God's word; some by the pleasure they take in the vanities of worldly things. But they that in this world live not after God, but after their own carnal liberty, perceive not this great wrath of God towards them, that he will not dig nor delve any more about them, that he doth let them alone even to themselves."[125]

Wesley's sermon, *A CALL TO BACKSLIDERS*, builds upon the biblical theme sounded in the HOMILY just cited. However, his treatment of the theme is found in many of his sermons. In a way, backsliding and fallen Christians were a preoccupation of Wesley in preaching, visitation, class meeting, and writing. In his *JOURNAL*, Wesley stated the basic idea of backsliding as a lack of upward movement on the ladder of personal salvation -

"It is impossible that any should retain what they receive, without improving it. Add to this, that the more we have received, the more care and labour is required, the more watchfulness and prayer, the more circumspection and earnestness in all manner of conversation. Is it any wonder, then, that they who forget this should soon lose what they had received?"[126]

In teleological salvation, only upward, heavenward movement is acceptable. There are no plateaus or promontories where one may rest. All is effort, all the time. To stop climbing is to fall back. Wesley's understanding of the process of salvation was clearly teleological. Hence backsliding, or declining from God, was frequently encountered. How will backsliders fare in the great judgment?

In the *CALL TO BACKSLIDERS*, Wesley answers this and many other questions. In the opening words of the sermon the gravity of declining from God is set forth - "Presumption is one grand snare of the devil, in which many of the children of men are taken. They so presume upon the mercy of God as utterly to forget his justice."[127] God has declared, "Without holiness no man shall see the Lord." Yet these presumers "flatter themselves, that in the end God will be better than his word." They "imagine they may live and die in their sins, and nevertheless escape the damnation of hell." These are backsliders, but so are those who despair of holiness, so that "they no longer strive; for they suppose it is impossible they should attain."[128] The former rush into sin, "as a horse into battle." They sin "with so high a hand, as utterly to quench the Holy Spirit of God; so that he gives them up to their own heart's lusts, and lets them follow their own imaginations."[129] These persons seem to be without fear of God, or hell. The god of this world blinds and hardens their hearts because they have chosen to sin. Wesley's view of sin is dynamic - no one sins because grace is not present. One who sins, sins in the presence of grace that could deliver from evil if appropriated. The sin of backsliding is, therefore, a tragic and unnecessary lapse from the active grace of God. Many a backslider acknowledges this tragedy, however, but despairs because he thinks that "he will be no more entreated."[130] Nevertheless, God has, in these last days, "visited again, and restored (some of them) to their first love."[131] This visitation is a reference by Wesley to the Methodist revival in which a major appeal was to backsliders to come back to God. For instance, Wesley's evangelical sermon on *THE MARKS OF THE NEW BIRTH* makes this point -

> "Say not then in your heart, 'I was once baptized, therefore I am now a child of God.' Alas, that consequence will by no means hold. How many are the baptized gluttons and drunkards, the baptized liars and common swearers, the baptized railers and evil-speakers, the baptized whoremongers, thieves, extortioners? What think you? Are these now the children of God? Verily, I say unto you, whosoever you are, unto whom any one of the preceding characters belongs, 'Ye are of your father the devil, and the works of your father ye do.' Unto you I call, in the name of Him whom you crucify afresh, and in his words to your circumcised predecessors, 'Ye serpents, ye generation of vipers, how can ye escape the damnation of hell?"[131]

A great many of the Methodists were reclaimed backsliders from the religion of the Church of England. And yet, Methodism produced many backsliders of its own. One notorious culprit was Westley Hall, one-time a devout Methodist, then a Deist, and finally an apostate. Of this case Wesley wrote -

> "About twelve years ago he was, without all question, filled with faith and the love of God. He was a pattern of humility, meekness, seriousness, and above all, of self-denial; so that, in all England, I knew not his fellow. It were easy to point out the several steps whereby he fell from his steadfastness; even till he fell into a course of adultery, yea, and avowed it in the face of the sun!"[132]

Wesley's sermon on backsliding attempts to assure such fallen persons that there is hope for restoration. The Bible, rightly understood, charts the path back to God.[133] The sum of the matter, according to Wesley, is -

> "None then can infer, that because an earthly king will not pardon one that rebels against him a second time, therefore the king of heaven will not. Yes, he will; not until seven times only, or until seventy times seven. Nay, were your rebellions multiplied as the stars of heaven; were they more in number

than the hairs of your head; yet, 'return unto the Lord, and he will have mercy upon you; and to your God, and he will abundantly pardon."[134]

By such preaching of God's mercy, Wesley did not mean to advocate habitual rebellion to God. His compassion for backsliders who felt hopelessly lost overwhelmed him until intellectual caution was temporarily minimized. Later on, this sermon warns that there is a "sin unto death," and backsliders may go so far as to commit it. Should they, their bodies will die unexpectedly as a punishment, so that their souls might be saved.[135] The mercy of God is given backsliders in this most remarkable situation, but no one should consider this as circumventing the justice of God. The body will be destroyed so that the soul may be corrected and survive the judgment into eternity. And yet, Wesley claimed, there are some backsliders who continue to flout all of God's grace and mercy, whose sin is not unto a premature death, and they refuse to come to repentance. Their continued declining from God, marked with obstinacy and rebellion, "only increases their damnation," which will be their sentence at the great judgment.[136] Lest any hearer might be tempted to presume upon God's mercy in reference to backsliding, Wesley closed the sermon with a solemn warning, implying that the judgment of a backslider might mean "the nethermost hell"

> "In all my experience, I have not known one who fortified himself in sin by a presumption that God would save him at the last, that was not miserably disappointed, and suffered to die in his sins. To turn the grace of God into an encouragement to sin is the sure way to the nethermost hell."[137]

As for backsliding Methodists, who do not get restored through repentance, and stand before the great throne of Christ's judgment, having failed to fulfill "all righteousness," they "will have the hottest place in the lake of fire," Wesley claimed. Indeed, Sodom and Gomorrah will fare better in the judgment than they![138] For Wesley the "nethermost hell" was the same as the "lake of fire."

For this reason - the possibility that salvation may be interrupted and even terminated along the way to judgment - Wesley believed that judgment must begin with the house of God, as St. Peter had affirmed (I Peter IV, 17). In Anglican theology of the late seventeenth century "final justification" was a popular doctrine, asserting that the great judgment either validates one's justified status or it does not. Bishop George Bull, in his famous *HARMONIA APOS-TOLICA*, stated the principle clearly -

"When we are justified in this life, a right to eternal life is truly conferred upon us, according to the law of Christ; when we are judged, in the next world, the same right is decided and confirmed by the solemn sentence of the Judge."[139]

According to Bishop Bull, final justification at the judgment is based on the law of Christ and not the law of Moses. The standard by which Christians shall be judged, therefore, is the gospel covenant of the New Testament in Christ.[140] The old law "gave no full and sufficient pardon for past sins, so neither did it afford any assistance to prevent future ones."[141] The new law, in Christ, however, does both, enabling "us to be dead unto sin itself, and alive unto God and true righteousness."[142] At the great judgment, Christians, who began to be dead to sin and alive to God in the mortal life, will be judged in reference to how they continued in the same through death. Should they have finished their course in faith, being dead to sin and alive to God, they shall be justified, finally and eternally. No one, Bull argued, should confuse this doctrine with the doctrine of the merit of works.[143] The two doctrines are diametrically opposite.

While Bishop Bull's understanding of final justification represents an Anglican perspective, Richard Baxter's shows a similar commitment from the Puritan side of the seventeenth century Church in England. In his *THE SAINTS' EVERLASTING REST*, Baxter states the case concisely - "(At) the public and solemn process at their (saints) judgment... they shall first themselves be acquitted and justified, and then with Christ judge the world."[144]

Wesley knew the Anglican and Puritan views of final justification. He disliked, however, the idea of two justifications, favoring the wording of Article XI of the Church, entitled *ON THE JUSTI-FICATION OF MAN*, in which but one justification is affirmed. Nevertheless, Wesley allowed the distinction -

> "The justification of which St. Paul and our Articles speak, is one only. And so say I still; and yet I do not deny that there is another justification (of which our Lord speaks) at the last day. I do not therefore condemn the distinction of a two-fold justification, in saying, That spoken of in our Articles is but one."[145]

Now that the outcome of judgment for one segment of humanity has been explored, the plight of the heathens in Wesley's thought is worth study. How will they fare in the great judgment? One very popular answer in Christian tradition has been - "Without a saving knowledge of Jesus Christ, they will be forever lost, damned in hell." This doctrine of *damnatio* rests upon the assumption that God wills to leave certain individuals in their corrupted states so they may inherit eternal punishment. Historically, there have been two versions of this: the supralapsarian position of "decrees," in which God elects whomever he wishes, apart from any consideration of the fall of man; and the infralapsarian position in which a negative act of God leaves some of the fallen in sin. Either way, the heathen are always assigned to hell following the great judgment. Wesley thought the doctrine itself, in both forms, was worthy of damnation. In his *PREFACE TO A TREATISE ON JUSTI-FICATION*, published November 16, 1764, Wesley launched an attack upon it -

> "What becomes of all other people? They must inevitably perish for ever.
> The die was cast or ever they were in being, The doctrine to pass them by has
> Consign'd their unborn souls to hell
> And damn'd them from their mother's womb!
> I could sooner be a Turk, a Deist, yea, an Atheist, than I could believe this. It is less absurd to deny the very being of God, than to make him an almighty tyrant."[146]

In a surprising manner, Wesley carried this attack a step further, making an affirmation that surely shocked his Reformed critics. In the middle of a sermon - number CXXV - *ON LIVING WITHOUT GOD* - he boldly affirmed -

"Let it be observed, I purposely add, to those that are under the Christian dispensation; because I have no authority from the word of God 'to judge those that are without;' nor do I conceive that any man living has a right to sentence all the heathen and Mahometan world to damnation. It is far better to leave them to Him that made them, and who is 'the Father of the spirits of all flesh;' who is the God of the heathens as well as the Christians, and who hateth nothing that he hath made. But, meantime, this is nothing to those that name the name of Christ; - all those, being under the law, the Christian law, shall undoubtedly be judged thereby; and, of consequence, unless those be so changed as was the animal above mentioned, unless they have new sense, ideas, passions, tempers, they are no Christians."[147]

This affirmation was not an emotional response in the heat of debate. On the contrary, it was calm and reasoned, having its roots in a theological system frequently called Arminianism. From that perspective, which entered the Anglican tradition in the seventeenth century, Wesley believed the grace of God comes in various stages in man's encounter with God. One particular kind of grace, reaching every person at birth, he called "preventing grace." This gift of grace is universal because the atonement of Christ was universal, that is, for every human, being "a full, perfect, and sufficient sacrifice for the sins of the whole world." Therefore, "there is no man that is in a state of mere nature," or without the grace of God.[148] This preventing grace is a supernatural gift at birth, in the form of "conscience." Some have more of it than others, but everyone has the gift. This conscience is likened to light - "some faint glimmering ray, which, sooner or later, more or less, enlightens every man that cometh into the world."[149] Wesley argued, "No man sins because he has not grace, but because he does not use the grace which he hath."[150] This is the beginning of salvation, and it is a common beginning for all men, everywhere and of every century -

"Salvation begins with what is usually termed (and very properly) *preventing grace*; including the first wish to please God, the first dawn of light concerning his will, and the first slight transient conviction of having sinned against him. All these imply some tendency toward life; some degree of salvation; the beginning of a deliverance from a blind, unfeeling heart, quite insensible of God and the things of God. Salvation is carried on by *convincing grace*, usually in Scripture termed *repentance*; which brings a larger measure of self-knowledge, and a farther deliverance from the heart of stone. Afterwards we experience the proper Christian salvation; whereby, 'through grace,' we are saved by faith, consisting of those two grand branches, justification and sanctification."[151]

Moreover, Wesley claimed, God's work of salvation is to "justify, to sanctify, and to glory" fallen man, and in some way, man must "work together with God."[152] But, some men are never exposed to the preaching of the gospel of Jesus Christ so that they come to faith in response to convincing and sanctifying grace. What must become of them in the great judgment? God, through preventing grace, continues to work in their lives, Wesley argued. God treats such heathens as a special dispensation, and he rewards those who diligently seek him.[153] Furthermore, according to Wesley, there are different types of heathens. There are (1) heathens of the baser sort -

"Many of them inferior to the beasts of the field. Whether they eat men or no, (which indeed I cannot find any sufficient ground to believe,) they certainly kill all that fall into their hands. They are, therefore, more savage than lions; who kill no more creatures than are necessary to satisfy their present hunger."[154]

And (2) there are what Wesley termed "enlightened heathens," who, both in ancient and present times, know nothing about the justification and sanctification that is offered in Jesus Christ. Nevertheless, they are enlightened in other ways.[155] Of these heathens Wesley spoke in his sermon entitled, *WALKING BY SIGHT AND BY FAITH* -

"Even the heathens did not all remain in total darkness... Some few rays of light have, in all ages and nations, gleamed through the shade. Some light they derived from various fountains touching the invisible world. 'The heavens declared the glory of God,' though not to their outward sight: 'The firmament showed,' to the eyes of their understanding, the existence of their Maker. From the creation they inferred the being of a Creator, powerful and wise, just and merciful. And hence they concluded, there must be an eternal world, a future state, to commence after the present; wherein the justice of God in punishing wicked men, and his mercy in rewarding the righteous, will be openly and undeniably displayed in the sight of intelligent creatures."[156]

In addition to these qualities of enlightened heathens, Wesley argued, many possessed a great degree of honesty, from whence they taught that -

"they ought not be unjust; not to take away their neighbour's goods, either by robbery or theft; not to oppress the poor, neither to use extortion toward any; not to cheat, or over-reach either the poor or rich, in whatsoever commerce they had with them; to defraud no man of his right; and, if it were possible, to owe no man anything."[157]

Likewise, enlightened heathens long have been concerned for truth and justice. They have held the perjurer to be an abomination, as well as the slanderer and false witness. Habitual liars they treated as the "disgrace of human kind, and the pests of society."[158] Furthermore, they prized love and assistance, expecting everyone to reciprocate the same toward others. They encouraged their kind to feed the hungry, "if they had food to spare." Also they encouraged those with superfluous clothing to share with those without.[159]

Wesley turned to St. Paul for a summary of what kind of a conscience rules an enlightened heathen. In his epistles, Paul makes the conscience, given as preventing grace, "a faculty or power, implanted by God in every soul that comes into the world, of perceiving what is right and wrong in his own heart or life, in his tempers, thoughts, words, and actions."[160] So, heathen honesty, truth, justice, love, and assisting those in need, all spring from this divine gift. Indeed, for Wesley, they do, and he called these principles "the rule of Heathens." Citing St. Paul, Wesley observed -

> "The rule of the Heathens, as the Apostle teaches elsewhere, is 'the law written in their hearts.' 'These,' saith he, 'not having the' outward 'law, are a law unto themselves: Who show the work of the law,' and which the outward law prescribes, 'written in their heart;' by the finger of God; 'their conscience also bearing witness,' whether they walk by this rule or not, 'and their thoughts the meanwhile accusing, or even excusing,' acquitting, defending them" (Romans II, 14-15).[161]

What may the enlightened heathens, under these conditions described by Wesley, find at the great judgment? Mercy and justice. During conversations with his preachers, this subject came up for discussion. A question was asked, "Who of us is now accepted of God?" The answer - "He that now believes in Christ with a loving, obedient heart." Another question was posed, "But who among those who never heard of Christ?" The answer, pertinent to this study, was given affirmatively - "He that, according to the light he has, 'feareth God and worketh righteousness.'"[162] In his *PREDESTINATION CALMLY CONSIDERED*, Wesley explained the ground for this affirmation -

> "Our God is just in all his ways; he reapeth not where he hath not strewed. He requireth only according to what he hath given; and when he hath given little, little is required. The glory of his justice is this, to 'reward every man according to his works.' Hereby is that glorious attribute shown, evidently set forth before men and angels, in that it is accepted of every man according to that he hath, and not according to that he hath not. This is that just decree which cannot pass, either in time, or in eternity."[163]

In the day of judgment, Wesley often mentioned, there will be surprises. "Shall not the very Heathen then 'rise up in judgment against this generation,'" Wesley asked, "and condemn it?" Then he answered his own question - "Yea, and not only the learned Heathen of Greece and Rome, but the savages of America."[164] In fact, Wesley argued, some heathens will fare better in the judgment than some Christians. In a blistering letter to John Glass, a defrocked Presbyterian minister, who publicly slandered Wesley under the pen-name of Palaemon, Wesley concluded - "God be merciful to thee a sinner; and show thee compassion, although thou has not any for thy fellow servants! Otherwise it will be more tolerable, I will not say for Seneca or Epictetus, but for Nero and Domitian, in the day of judgment."[165] In his *JOURNAL*, Wesley showed a fascination with Emperor Marcus Aurelius (A.D. 121-180) whom he regarded as an enlightened heathen. Wesley was convinced that Aurelius would fare well in the judgment -

> "I make no doubt but this is one of those many who shall come from the east and west and sit down with Abraham, Isaac, and Jacob while the children of the kingdom, nominal Christians, are shut out."[166]

Wesley's optimistic prognostication of the lot of Aurelius in the judgment was expressed at the height of his evangelical activities, in 1745. No one could accuse him then of being old, senile and sentimental about the state of the heathens. As early as 1733, however, he was vitally concerned with the spiritual state of heathens everywhere, and in his published prayers for children, he uttered the petition - "Prosper all those who are sincerely engaged in propagating or promoting thy faith and love, - Give thy Son the Heathen for his inheritance, and the utmost parts of the earth for his possession; that from the rising up of the sun unto the going down of the same, thy name may be great among the Gentile."[167] Two years later, he went to Georgia as a missionary to the heathens there, and he found more heathens among the colonists than he found among the savages.

Notes

1   METHODIST HYMNAL, *op.cit.*, 631.
2   WORKS, *op.cit.*, VII, 325.
3   *Ibid*, VII, 324.
4   EXPLANATORY NOTES NEW TESTAMENT, *op.cit.*, 1050.
5   *Ibid*, 50, 147, 221, 314.
6   *Ibid*, 719.
7   *Ibid*, 117, 250.
8   *Ibid*, 118-119, 1033-1034.
9   *Loc.cit.*
10 *Ibid*, 116, 184.
11 WORKS, *op.cit.*, VIII, 350; XI, 408. Cf. JOURNAL, *op.cit.*, V, 4.
12 JOURNAL, *op.cit.*, V, 9.
13 *Loc.cit.*
14 Southey, *op.cit.*, II, 344-345.
15 WORKS, *op.cit.*, XII, 319.
16 EXPLANATORY NOTES NEW TESTAMENT, *op.cit.*, 116.
17 *Ibid*, 898.
18 *Ibid*, 761.
19 *Ibid*, 898.
20 *Ibid*, 760.
21 *Ibid*, 937.
22 *Ibid*, 84.
23 *Ibid*, 95.
24 *Ibid*, 115.
25 *Ibid*, 1035-1039.
26 *Ibid*, 1038-1039.
27 *Ibid*, 1041.
28 *Ibid*, 464.
29 WORKS, *op.cit.*, VII, 474.
30 EXPLANATORY NOTES NEW TESTAMENT, *op.cit.*, 632.
31 WORKS, *op.cit.*, VII, 474.
32 *Ibid*, VII, 474-475.
33 *Ibid*, VII, 475.
34 *Ibid*, VII, 476.
35 *Loc.cit.*
36 *Loc.cit.*
37 *Ibid*, VII, 477.
38 *Loc.cit.*
39 *Loc.cit.*

40 *Loc.cit.*
41 *Ibid*, VII, 478.
42 *Loc.cit.*
43 *Ibid*, VII, 479.
44 *Loc.cit.*
45 *Loc.cit.* Cf. NICENE FATHERS, *op.cit.*, III, 266. St. Augustine believed that the resurrected bodies of the wicked will retain previous deformities in their immortal and incorruptible state.
46 *Loc.cit.*
47 *Ibid*, VII, 480.
48 *Loc.cit.*
49 *Loc.cit.*
50 *Loc.cit.*
51 *Ibid*, VII, 481.
52 *Loc.cit.*
53 *Loc.cit.*
54 *Ibid*, VII, 482.
55 *Loc.cit.*
56 *Loc.cit.*
57 *Loc.cit.*
58 *Loc.cit.*
59 *Ibid*, VII, 482-483.
60 *Ibid*, VII, 483.
61 *Loc.cit.*
62 *Loc.cit.*
63 *Ibid*, VII, 484.
64 *Loc.cit.*
65 Norman Cohn, THE PURSUIT OF THE MILLENNIUM (Fairlawn: Essential Books, 1957), 21.
66 EXPLANATORY NOTES NEW TESTAMENT, *op.cit.*, 1036-1037.
67 *Ibid*, 1038.
68 *Ibid*, 1039.
69 *Loc.cit.*
70 *Loc.cit.*
71 *Ibid*, 1051. Cf. WORKS, *op.cit.*, XII, 437. When Wesley's view of A.D. 1836 became well-known, he disowned it and blamed Bengelius whom he had quoted.
72 WORKS, *op.cit.*, X, 31.
73 Philip Schaff, THE ANTI-NICENE FATHERS (Buffalo: The Christian Literature Company, 1887), I, 240.
74 Tyerman, *op.cit.*, II, 524.
75 Telford, *op.cit.*, IV, 197-198.
76 WORKS, *op.cit.*, XII, 244. Cf. Telford, *op.cit.*, IV, 234.
77 THE ARMINIAN MAGAZINE (London: J. Paramour, 1784), III, 155.
78 *Ibid*, III, 154.
79 *Ibid*, III, 209.

80   WORKS, *op.cit.*, VI, 279.
81   *Loc.cit.*
82   *Ibid*, VI, 280.
83   *Ibid*, VI, 281.
84   *Ibid*, VI, 283.
85   *Loc.cit.*
86   *Ibid*, VI, 284.
87   *Loc.cit.*
88   *Ibid*, VI, 285.
89   *Loc.cit.*
90   *Ibid*, VI, 286.
91   *Loc.cit.*
92   *Ibid*, VI, 286-287.
93   *Ibid*, VI, 287.
94   *Ibid*, VI, 288.
95   JOURNAL, *op.cit.*, IV, 254.
96   WORKS, *op.cit.*, V, 172.
97   *Ibid*, V, 173.
98   *Loc.cit.*
99   *Loc.cit.*
100   *Ibid*, V, 174.
101   NICENE FATHERS, *op.cit.*, II, 445.
102   WORKS, *op.cit.*, V, 175. Cf. EXPLANATORY NOTES NEW TESTA-
      MENT, *op.cit.*, 760. "The wicked will remain beneath, while the righteous,
      being absolved, shall be assessors with their Lord in judgment."
103   *Loc.cit.*
104   *Loc.cit.*
105   *Ibid*, V, 176.
106   *Loc.cit.*
107   *Loc.cit.*
108   *Ibid*, V, 177.
109   *Loc.cit.*
110   *Ibid*, V, 178.
111   *Loc.cit.*
112   *Loc.cit.*
113   *Loc.cit.*
114   *Ibid*, V, 179.
115   *Loc.cit.*
116   *Ibid*, V, 181.
117   ARMINIAN MAGAZINE, *op.cit.*, V, 674.
118   WORKS, *op.cit.*, X, 320ff; XII, 208, 283, 290.
119   *Ibid*, VI, 509.
120   Naglee, *op.cit.*, 38-42.
121   WORKS, *op.cit.*, X, 363.
122   EXPLANATORY NOTES NEW TESTAMENT, *op.cit.*, 810.
123   *Ibid*, 897.

124 HOMILIES, (1833), *op.cit.*, 53.
125 *Ibid*, 56-57.
126 JOURNAL, *op.cit.*, V, 106.
127 WORKS, *op.cit.*, VI, 514.
128 *Loc.cit.*
129 *Ibid*, VI, 515.
130 *Loc.cit.*
131 *Ibid*, V, 221-222.
132 JOURNAL, *op.cit.*, III, 324.
133 WORKS, *op.cit.*, VI, 517-519.
134 *Ibid*, VI, 519.
135 *Ibid*, VI, 520. Cf. V, 83. Here Wesley contradicts his sermon on backsliding by claiming the "sin unto death" means a "never-ending death with everlasting destruction" - this was St. Augustine's position.
136 *Ibid*, VI, 525.
137 *Ibid*, VI, 526-527.
138 JOURNAL, *op.cit.*, V, 240.
139 George Bull, *HARMONICA APOSTOLICA* (Oxford: J.H. Parker, 1842), 30.
140 *Ibid*, 94.
141 *Ibid*, 95.
142 *Ibid*, 96.
143 *Ibid*, 207.
144 Richard Baxter, THE SAINTS EVERLASTING REST (11th Edition, Soho: J. Barfield, 1817), 24.
145 WORKS, *op.cit.*, X, 430-431.
146 *Ibid*, X, 334.
147 *Ibid*, VII, 353.
148 *Ibid*, VI, 512.
149 *Loc.cit.*
150 *Loc.cit.*
151 *Ibid*, VI, 509.
152 *Ibid*, X, 230.
153 *Ibid*, VII, 195.
154 *Ibid*, VI, 278.
155 *Ibid*, VI, 506.
156 *Ibid*, VII, 258. Cf. V, 8; IX, 388.
157 *Ibid*, V, 17.
158 *Ibid*, V, 17-18.
159 *Ibid*, V, 18.
160 *Ibid*, V, 136.
161 *Loc.cit.*
162 *Ibid*, VIII, 337.
163 *Ibid*, X, 234. Cf. Southey, *op.cit.*, II, 76f.
164 *Ibid*, VIII, 150.
165 Telford, *op.cit.*, III, 239.

166	JOURNAL, *op.cit.*, III, 215.
167	WORKS, *op.cit.*, XI, 205.

Chapter Twelve

Paradise Improved: The New Creation

"There is my house and portion fair;
My treasure and my heart are there,
And my abiding home;
For me my elder brethren stay,
And angels beckon me away,
And Jesus bids me come.

'I come,' thy servant, Lord, replies,
'I come to meet thee in the skies,
And claim my heavenly rest!
Now let the pilgrim's journey end;
Now, O my Savior, Brother, Friend,
Receive me to thy breast!'"[1]

At long last the pilgrim's journey ends, as Wesley's hymn claims. But, he is not the only pilgrim struggling through time. The heavens and the earth, and all that is within, are pilgrims too. Christian pilgrims, as St. Paul declares, reckon that "the sufferings of the present time are not worthy to be compared with the glory which shall be revealed" (Romans VIII, 18). Nevertheless, every pilgrim suffers and often groans with sounds that are unintelligible. The hope in a "glory which shall be revealed" is a genuine longing for an audience with God, face to face, in which He shall wipe away all tears, shall banish all sorrow and sighing, and remove all pain. St. Augustine, in his *CIVITAS DEI*, spoke of this experience -

"The holier a man is, and the fuller of holy desire, so much the more abundant is the tearfulness of his supplication. . . Are not these the utterances of a citizen of the heavenly Jerusalem (the Church): 'My tears have been my meat day and night;' and 'Every night shall I make my bed to swim; with my tears shall I water my couch;' and 'My sorrow was renewed?' Or are not

those God's children who groan, being burdened, not that they wish to be un-
clothed, but clothed upon, that mortality may be swallowed up of life? Do
not they even who have the first-fruits of the Spirit groan within themselves,
waiting for the adoption, the redemption of their body?"[2]

Wesley agreed with both Paul and Augustine in describing the
longing of every true pilgrim, groaning for mortality to be swallowed up in eternal life. Immortal, incorruptible, eternal life, for
Wesley, is the *telos* of the pilgrim's salvation, begun on earth and in
time but completed and perfected in heaven,[3] after the great judgment has passed. However, St. Paul added the entire creation to
the pilgrim's suffering and hope for final redemption -

"The creation was made subject to vanity, not willingly, but by him who subjected it. In hope that the creation itself shall be delivered from the bondage
of corruption into the glorious liberty of the children of God. For we know
that the whole creation groaneth together and travaileth together until now"
(Romans VIII, 20-22).

In his commentary on the suffering and groaning creation, Wesley stated -

"*Of the creation* - Of all visible creatures, believers excepted, who are spoken
of apart; each kind, according as it is capable. All these have been sufferers
through sin; and to all these (the finally impenitent excepted) shall refresh-
ment redound from the glory of the children of God. Upright heathens are
by no means to be excluded from this *earnest expectation*."[4]

This personal and cosmic redemption, for pilgrim and creation
alike, is the stuff of which good sermons are made. Wesley did not
let the occasion to preach on this subject pass by. He chose to call
the potent sermon, numbered LX, *THE GENERAL DELIVER-
ANCE*, and its biblical text was Romans VIII, 19-22. The sermon
begins with a scriptural account of God's beneficence to "every
man" and "all his works." The main body of the sermon develops
three points - (1) the original state of the brute creation; (2) Its
present state; and (3) What its state shall be at the manifestation of

the children of God.[5] The homily proceeds chiefly along biblical lines, using reason for analysis, treating the nature of man and the lesser creatures. As this chapter progresses, the sermon will be cited several times. But for now, let it suffice to recognize that time, in Wesley's theology, has reached its end, and eternity *a parte post* has immediately begun. In this end which is the beginning, there are three dimensions of God's consummation of all things in Christ - The old passes away, all things become new, and the new Jerusalem is established.

THE OLD PASSES AWAY

Wesley rightly understood that the Apocalypse is the primary source for the doctrine of time being swallowed up by eternity. Therefore, his commentary on chapters XX through XXII of that book are foundational to his numerous affirmations concerning this subject. The great judgment (XX, 11-15) is the dividing line where time ceases and eternity begins for God's creation. The righteous, in their resurrected bodies, enter into the very life of God. The "finally impenitent," in their resurrected bodies, are cast into the lake of fire, where Satan, Gog and Magog, previously had been thrown.[6] The Apostle John, Wesley noted, placed the "dissolution" of the created heavens and the earth in the context of the great judgment.[7] Such a "removal" of these parts of the creation, he argued, does not constitute destruction, but being "wholly dissolved, the very 'elements melting with fervent heat.'"[8] For a more explicit statement of this melting dissolution, Wesley turned to the writing of St. Peter, whom he believed was the author of II Peter. His commentary observations on II Peter III, 10ff, help to make explicit what is only implicit in Revelation XX. "The day of the Lord" is the lengthy judgment of at least a thousand years; and accompanying it, probably at the end of it, "the heavens shall pass away with a great noise," and, moreover, "the elements shall melt with fervent heat, and the earth and the works therein shall be burned up." Wesley's treatment follows -

"The elements shall melt with fervent heat - The elements seem to mean, the sun, moon, and stars; not the four commonly so called; for air and water cannot melt; and the earth is mentioned immediately after. *The earth and all the works* - Whether of nature or art. . . Has not God already abundantly provided for this? (1) By the stores of subterranean fire which are so frequently bursting out at Aetna, Vesuvius, Hecla, and many other burning mountains. (2) By the ethereal (vulgarly called electrical) fire, diffused through the whole globe; which, if the secret chain that now binds it up were loosed, would immediately dissolve the whole frame of nature. (3) By comets, one of which, if it touch the earth in its course toward the sun, must needs strike it into that abyss of fire; if in its return from the sun, when it is heated, as a great man computes, two thousand times hotter than a red-hot cannon-ball, it must destroy all vegetables and animals long before their contact, and soon after burn it up."[9]

Following Bengelius, Wesley argued that the dissolution of the heavens and the earth are possible by virtue of God having written into nature the means for accomplishing it - volcanic eruptions, lightning, and comets. In no way should the future passing away of heaven and earth be regarded as evil, a work of Satan (who by this time is already in the eternal and nethermost hell), nor the work of man with twentieth century nuclear arms. It is entirely the work of God, for Wesley. One wonders why Wesley did not apply Hebrews XII, 29 to this motif - "For our God is a consuming fire!" His exegesis of the verse reads - "In the strictness of His justice, and purity of His holiness."[10] Such an application would affirm God purifying the heavens and the earth by his own nature as fire, using volcanic activity, lightning, or comets heated by the sun, to bring to reality his justice and holiness. This emphasis would make solid Wesleyan dogma. It should be observed, moreover, that Wesley did precisely this, but in other words, words from the *GREAT ASSIZE* sermon -

"Then the heavens will be shrivelled up as a parchment scroll, and pass away with a great noise: They will 'flee from the face of him that sitteth upon the throne, and there will be found no place for them.' (Rev. XX, 11) The very manner of their passing away is disclosed to us by the Apostle Peter: 'In the day of God, the heavens, being on fire, shall be dissolved.' (II Peter III, 12) The whole beautiful fabric will be overthrown by that raging element, the connexion of all its parts destroyed, and every atom torn assunder from the others. By the same, 'the earth also, and the works that are therein, shall be burned up.' (Verse 10) The enormous works of nature, the everlasting hills, mountains that have defied the rage of time, and stood unmoved so many thousand years, will sink down in fiery ruin. How much less will the works of art, though of the most durable kind, the utmost efforts of human industry, - tombs, pillars, triumphal arches, castles, pyramids, - be able to withstand the flaming conqueror! All, all will die, perish, vanish away, like a dream when one awaketh!"[11]

We must not overlook a significant phrase used by Wesley above - "every atom torn assunder from the others." As noted in Chapter III (The Creation), Wesley was a Christian atomist, believing that atoms are the basic building blocks of material objects. Although created, God gave them an eternal dimension, so they are eternal and indestructible. As in the atomism of ancient Democritus, Wesley's atoms exist in relationships with other atoms to comprise a thing of matter. When they "split off" from such relationships, those material things pass out of being, and yet the atoms continue to exist. This was important to Wesley in reference to the heavens and the earth passing away - things composed of atoms shall cease to be since their atoms will separate from another. But, most importantly, the atoms will not be destroyed. God will give them new relationships from which will come "a new heaven and a new earth," pure, clean, innocent, and holy, created for God and eternity - the "New" being superior to the "Old" in every way. The "Old" was made good, at the beginning of time, but not perfect. The "Old" was to go on to perfection over the duration of time, but sin entered and cursed, in some degree, that plan. God's grace countered the flaw of sin, abounding but never completely destroying its power. Now the purification of God, at the end of time, removes the cosmic curse, and a "New" creation, built with the atoms from the first creation, takes its eternal place - *a parte post*!

ALL THINGS BECOME NEW

John, the prophet of Patmos, recorded the vision - "And I saw a new heaven and a new earth: for the first heaven and the first earth were passed away; and there was no more sea" (Rev. XXI, 1). Wesley argued that this vision "reaches into eternity." Indeed it does, in his system. He was insistent that the passing of the old and the coming of the new follow the resurrection and general judgment.[12] Taking exception to a popular Roman Catholic interpretation, that the "new creation" is the Church in a flourishing state, Wesley claimed it is "a new eternal state of things."[13] While his *EXPLANATORY NOTES UPON THE NEW TESTAMENT* states the case clearly and concisely, it is his sermon, *THE NEW CREATION*, that gives us the depth of Wesley's understanding of God's consummate act of power.

Using Revelation XXI, 5 as the text - "Behold, I make all things new" - the words of the great judge and king, Jesus Christ - Wesley started the sermon by trying to rescue the passage from the faulty interpretation cited above -

> "Very many commentators entertain a strange opinion, that this relates only to the present state of things; and gravely tell us, that the words are to be referred to the flourishing state of the Church which commenced after the heathen persecutions. Nay, some of them have discovered, that all which the Apostle speaks concerning the 'new heaven and the new earth' was fulfilled when Constantine the Great poured in riches and honours upon the Christians. What a miserable way is this of making void the whole counsel of God, with regard to that grand chain of events in reference to his Church, yea, and to all mankind, from the time that John was in Patmos, unto the end of the world! Nay, the line of this prophecy reaches further still: It does not end with the present world, but shows us the things that will come to pass, when this world is no more."[14]

Christ, in this sermon, is identified as both "Creator and Governor." He not only creates, but he governs. In his infinite power and wisdom, he can cause his creation to be dissolved so that he can make a new one from the indestructible particles of the former creation. Hence, he announces, "Behold, I make all things new."

For the Apostle, Wesley said, "all things" meant "a new heaven and a new earth." Here Wesley treats the subject of "multiple heavens" in the Hebrew and Greek Testaments of the Bible, a subject, when treated before a congregation, brings skepticism from some and indifference from others. Woe unto the lecturer who mentions Hebrew names for at least seven heavens - *Vilon, Rakia, Sheshakim, Zebul, Maon, Machon,* and *Araboth,* for his audience will either bristle with hostility or fall asleep on him. In many congregations, a lecturer in New Testament theology may be added to the "prayer list" after he mentions that in Matthew's gospel the Greek reads "the kingdom of heavens" and not "the kingdom of heaven." Be this as it may, Wesley addressed the subject head-on -

> "*A new heaven*: The original word in Genesis (chapter I) is in the plural number: and indeed, this is the constant language of Scripture; not *heaven*, but *heavens*. Accordingly, the ancient Jewish writers are accustomed to reckon three heavens; in conformity to which, the Apostle Paul speaks of his being caught 'up into the third heaven.' It is this, the third heaven, which is usually supposed to be the more immediate residence of God; so far as any residence can be ascribed to his omnipresent Spirit, who pervades and fills the whole universe. It is here (if we speak after the manner of men) that the Lord sitteth upon his throne, surrounded by angels and archangels, and by all his flaming ministers."[15]

The third heaven, Wesley argued, being uncreated, is not to be involved in the passing away of the heavens. "Only the interior heavens are liable to change, the highest of which we usually call the starry heavens." The Apostle Peter said that this heaven is "reserved unto fire, against the day of judgment and destruction of ungodly men."[16] Sometime during the latter part of the judgment, after the righteous have been justified, and while the wicked are before the great judge, "being on fire, it shall, first, shrivel as a parchment scroll; then it shall be dissolved, and shall pass away with a great noise; lastly, it shall flee from the face of Him that sitteth on the throne, and there shall be found no place for it."[17] Wesley continued, having, at the same time, the stars falling from their former places - "the secret chain being broken which had retained them in their several orbits from the foundation of the world."[18] Then the lower or "sublunary" heaven, that contains the elements that compose it, shall begin to burn with fervent heat. The earth then shall begin to burn, and all the works of it shall be burned up. Wesley's conclusion was optimistic - "This is the introduction to a far nobler state of things, such as it has not yet entered into the heart of men to conceive, - the universal restoration, which is to succeed the universal destruction." An apostolic anticipation is strategically stated by Wesley - "For we look for new heavens and a new earth, wherein dwelleth righteousness" (II Peter III, 7).[19]

The sermon treats the "new heavens" by explaining that there will be no blazing stars within them, no comets. All will be "exact order and harmony" in the new heavens.[20] As for the lowest heaven of air, that too will be different. No hurricane nor furious storms shall arise within it. "Pernicious or terrifying meteors will have not place therein." Concerning the new heavens, "all will be light, fair, serene; a lively picture of the eternal day."[21]

The "elements" of the created order demand a definition under the new conditions. Wesley's Aristotelian physics are clearly seen here. "All the elements... will be new indeed; entirely changed as to their qualities (attributes), although not as to their nature (substance)." For instance, presently fire is a destroyer, dissolving things "that come within its sphere of action." Fire reduces material objects to their "primitive atoms." In the new order of things, fire will behave differently -

"But no sooner will it have performed its last great office of destroying the heavens and the earth; (whether you mean thereby one system only, or the whole fabric of the universe; the difference between one and millions of worlds being nothing before the great Creator;) when, I say, it has done this, the destructions wrought by fire will come to a perpetual end. It will destroy no more: It will consume no more: It will forget its power to burn, - which it possesses only during the present state of things, - and be as harmless in the new heavens and earth as it is now in the bodies of men and other animals, and the substance of trees and flowers, in all which (as late experiments show) large quantities of ethereal fire are lodged; if it be not rather an essen-

tial component part of every material being under the sun. But it will probably retain its vivifying power, though divested of its power to destroy."[22]

One must remember Wesley's physics as being a blend of atoms
and the four ancient elements of air, fire, water, and earth. The elements are blended together in various proportions in each and
every material thing. Hence, fire is to be found in flowers as well as
in animals, together with the other elements in their several proportions. Wesley's sense of what is good physics could be support-
ive of biblical truth. The only problem with Wesley's physics is that
it is ancient and, since the time of Leibniz and his units of force,
terribly out of scientific respectability.

Because of the new heavens, there will be no more rain, Wesley
claimed, which was the case originally in the first Paradise of creation (Genesis II, 5-6). Furthermore, there will be no clouds or
fogs, "but one bright refulgent day." Nor will there be any
"poisonous damps" or "pestilential blasts." There will be no
"Sirocco" in Italy, and "no parching and suffocating winds in Arabia."[23]

The element of water shall also undergo a radical change in the
new order, Wesley argued. In the new, it will be "clear and limpid,"
being "pure from all unpleasing or unhealthful mixtures." It will
spring up as crystal fountains everywhere, and "adorn the earth
with liquid lapse of murmuring stream." As in the original Paradise,
so in the new, "there will be various rivers gently gliding along, for
the use and pleasure of both man and beast." But, "there will be no
more sea." For Wesley this meant that the seas will retreat back
into the bowels of the earth, where they were placed at creation.[24]
As a result, the surface of the new earth will be greater than in the
former. The food-producing qualities of the earth shall then be
multiplied greatly. But Wesley, at this point, strains the argument
with an affirmation that the inhabitants of the new earth "shall be
equal to angels." He did not mean "in every way like angels," requiring no food, etc., but rather, that humans shall be on "a level
with them in swiftness, as well as strength; so that they can, quick as

thought, transport themselves, or whatever they want, from one side of the globe to the other."[25]

Wesley's sermon continued with a treatment of the changes wrought in "earth" as an element. Citing the concept of Jacob Behme, a German mystic, that in the new order the things of earth shall be "transparent as glass," Wesley disclaimed the notion as being without scriptural or rational value. The visible features of the human body, he reasoned, communicates beauty that transparency cannot embellish. As for the earth itself, the extremes of hot and cold shall no longer rake it. The climate shall become temperate and "conducive to its fruitfulness."[26] In so far as internal phenomena are concerned, the earth shall no longer be convulsed by quakes. Moreover, its surface will be devoid of horrid rocks, wild deserts, barren sands, impassable morasses, or unfruitful bogs "to swallow up the unwary traveller." Gently rising hills will abound as ornaments of beauty. The "general produce of the earth" will not be "thorns, briers, or thistles; nor any useless or fetid weed; not any poisonous, hurtful, or unpleasant plant; but every one that can be conducive, in any way, either to our use or pleasure." Above all, Wesley claimed, "the earth shall be a more beautiful Paradise than Adam ever saw" - Paradise Improved![27]

The sermon moves from consideration of inanimate things being improved to animate things. Wesley's description is best -

"In the living part of the creation were seen the most deplorable effects of Adam's apostasy. The whole animated creation, whatever has life, from leviathan to the smallest mite, was thereby made subject to such vanity, as the inanimate creatures could not be. They were subject to that fell monster, DEATH, the conqueror of all that breathe. They were made subject to its fore-runner, pain, in its ten thousand forms; although God made not death, neither hath he pleasure in the death of any living. . . .He that sitteth upon the throne will soon change the face of all things, and give a demonstrative proof to all his creatures, that 'his mercy is over all his works.' The horrid state of things which at present obtains, will soon be at an end. On the new earth, no creature will kill, or hurt, or give pain to any other. The scorpion will have no poisonous sting; the adder, no venomous teeth. The lion will have no claws to tear the lamb; no teeth to grind his flesh and bones. Nay, no creature, no beast, bird, or fish, will have any inclination to hurt any other; for cruelty will be far away, and savageness and fierceness be forgotten. So that violence shall be heard no more, neither wasting or destruction seen on the face of the earth."[28]

Wesley's thesis is fascinating - When the Creator creates the "new earth" from the atoms and elements of the former, he shall make a new animal kingdom and spread it throughout the new earth. These creatures will be made immortal, and they shall never know pain. They shall not groan under suffering.

In his sermon, *THE GENERAL DELIVERANCE*, Wesley gave a protracted treatment of these lesser creatures and their place in the new earth. His argument ran - When created, the brute creatures were blessed by God through Adam, the channel of God's communication to them. Adam's sin broke that communication, and the animals and other lesser creatures were deprived of God's blessing, although his mercy continued their existence.[29] This condition is what St. Paul referred to as being "subjected to vanity," but these creatures also suffered the loss of some of their faculties (the serpent lost his speech). In addition, they lost some of their natural "vigour, strength, and swiftness." But the greatest loss suffered by these creatures was in their understanding, in their wills, in their passions, and in their natural love for man. Hence, "savage fierceness" with "unrelenting cruelty" possessed them, and they began to feed upon one another.[30] Wesley asked - "Where is the beauty which was stamped upon them when they came first out of the hands of their Creator?" The question is reminiscent of a Thomistic description of God's creative acts - "He made angels to show Him splendor; He made animals to show Him beauty; and He made man to serve Him out of the tangle of his mind." But, alas, man's sin tangled the various faculties of creaturely life, bringing pain and death. As for "beauty" - "There is not the least trace of it left: So far from it, that they are shocking to behold!" They are grisly and terrible in their present appearance, "deformed, and that to a high degree."[31] Pain, coming upon them as a result of Adam's sin, distorted their features. Pain was caused by weakness, sickness, diseases, from one another, inclemency of the seasons, and "from a thousand causes which they cannot foresee or prevent."[32] Death passed from Adam onto them, making man their "common enemy."[33]

Wesley's vivid account of the suffering of the brute creation turns optimistic -

> "But will 'the creature,' will even the brute creation, always remain in this deplorable condition? God forbid that we should affirm this; yea, or even entertain such a thought!   While 'the whole creation groaneth together,' (whether men attend or not) their groans are not dispersed in idle air, but enter into the ears of Him that made them. While his creatures 'travail together in pain,' he knoweth all their pain, and is bringing them nearer and nearer to the birth, which shall be accomplished in its season. He seeth 'the earnest expectation' wherewith the whole animated creation 'waiteth for' that final 'manifestation of the sons of God;' in which, 'they themselves also shall be delivered' (not by annihilation; annihilation is not deliverance) 'from the' present 'bondage of corruption into' a measure of 'the glorious liberty of the children of God.'"[34]

What does this mean? "Inclusion in the new earth," answered John Wesley. More specifically -

> "A general view of this is given us in the twenty-first chapter of the Revelation. When He that 'sitteth upon the great white throne' hath pronounced, 'Behold, I make all things new; when the word is fulfilled. 'The tabernacle of God is with men, and they shall be his people, and God himself shall be with them and be their God;' - Then the following blessing shall take place (not only on the children of men; there is no such restriction in the text; but) on every creature according to its capacity: 'God shall wipe away all tears from their eyes. And there shall be no more death, neither sorrow, nor crying. Neither shall there be any more pain: For the former things are passed away.'"[35]

Moreover, the entire brute creation will then be restored, "not only to the vigour, strength, and swiftness which they had at their creation, but to a far higher degree of each than they ever enjoyed." Paradise for them is more than restored - it is improved!

They have now entered eternity - "the days of the groaning are ended!"[36]

And what of the righteous children of God? What will be their lot on the new earth? Wesley treated these questions in reference to the theme of the New Jerusalem, describing the city and the blessedness of those who are worthy to enter into its gates. Our discussion of this facet is now appropriate.

THE NEW JERUSALEM

The last vision of the Apocalypse relates the establishment of an eternal and heavenly city - the New Jerusalem - upon the new earth. Historically, this has been a popular vision, interpreted differently from time to time. For St. Augustine, the Church was the New Jerusalem.[37] For Jan Matthys, in the sixteenth century, the Anabaptist kingdom of Munster was the New Jerusalem on earth.[38] To some English Puritans, in the early seventeenth century, their colony in the New World was the New Jerusalem.[39] To John Wesley, however, the New Jerusalem was what the prophet John said it was, the city of God come down onto a new earth. Wesley took the passage - Revelation XXI, 1 - XXII, 5 - quite literally. He believed that the New Jerusalem, as described by the Apostle John, is "wholly new, belonging not to this world, nor to the millennium, but to eternity."[40] This position clearly rules out those of Augustine, Matthys, and New England Puritanism. That verse 4 (chapter XXI) claims there will be no more death, Wesley argued, indicates that "this whole description belongs not to time, but eternity."[41]

There are six major divisions in the New Jerusalem passage of Revelation - (1) XXI, 1-4; (2) XXI, 5-8; (3) XXI, 9-14; (4) XXI, 15-21; (5) XXI, 22-27; and (6) XXII, 1-5. An examination of the text and Wesley's commentary on each will prove beneficial to this study.

(1) XXI, 1-4 -

"And I saw a new heaven and a new earth: for the first heaven and the first earth were passed away; and there was no more sea. And I saw the holy city, the new Jerusalem, coming down from God out of heaven, prepared as a bride adorned for her husband. And I heard a loud voice out of heaven, saying, Behold, the tabernacle of God with men, and he will pitch his tent with them, and they shall be his people, and God himself shall be with them, and be their God. And he shall wipe away all tears from their eyes; and death shall be no more, neither sorrow, or crying, or pain be any more: because the former things are gone away."

Wesley noted that there is a succession of visions, starting in Revelation XIX, 11; moving to XX, 1, 4, 11; and finally to XXI, 1. The last vision, of course, is the culmination of all previous visions, form the time of Enoch to John on Patmos. It is this climatic vision that reveals time being transformed into eternity, and God's creatures being brought into his immediate presence for the eternal, beatific life. It is after the resurrection and the judgment that the new Jerusalem comes through the new heaven and onto the new earth. The former heavens and earth have passed away, leaving empty space to be filled with a "new heaven" and a "new earth." Wesley's idea here is obvious to all of philosophical background - if space is full, there can be no movement. The space once occupied by the former heavens and earth must be made empty before a new heaven and earth can take their places.[42] The new heaven, the new earth, and the new Jerusalem are inextricably linked, Wesley maintained, although the holy city was long in being prepared - "The patriarchs had a revelation and a promise of eternal glory in heaven. . . seeing he hath prepared for them a city."[43] Wesley disappoints us by not treating the phrase about the holy city "adorned as a bride for her husband." He surely knew the biblical tradition of a heavily veiled bride, clean and pure, wearing perfume and white garments, being hand delivered by her father to her husband so that cohabitation could follow, the two becoming as one flesh. It is a powerful analogy, showing the gift of the eternal Father of his daughter, new Jerusalem, to her husband, the faithful saints of all ages, so that they might cohabit forever on the new earth. However, Wesley did make the point that God is within the city, so that cohabitation with his saints is achieved and the covenant (*ketubah* - marriage contract?) is at last fulfilled.[44] God's loving care is shown in verse 4, where he wipes away all tears. "Under the former heaven, and upon the former earth, there were death and sorrow, crying and pain; all which occasioned many tears; but now pain and sorrow are fled away, and the saints have everlasting life and joy."[45]

(2) XXI, 5-8 -

"And he that sat upon the throne said, Behold, I make all things new. And he saith to me, Write: these sayings are faithful and true. And he said to me, It is done. I am the Alpha and the Omega, the beginning and the end. I will give to him that thirsteth of the fountain of the water of life freely. He that overcometh shall inherit these things; and I will be to him a God, and he shall be to me a son. But the fearful, and inbelieving, and abominable, and murderers, and whoremongers, and sorcerers, and idolaters, and all liars, their part is in the lake that burneth with fire and brimstone; which is the second death."

The saying of Christ, from his great white throne, was spoken to the entire retinue of faithful creatures, both of heaven and earth, and not just to St. John, Wesley asserted - "Behold, I make all things new!" The Christ who created in the beginning as Alpha, now in the end (*telos*) creates anew as Omega. In the first creation he created all things good. In the new creation he creates all things perfectly. The same atoms and elements used in the former creation, Christ uses for the new creation. Yet, he is the same Creator and Christ, Alpha and Omega, the beginning and end. "These sayings are faithful and true," he declares. Wesley's observation concerning the apostle's response to the speaker of these sayings is - "The apostle seems again to have ceased writing, being overcome with ecstasy at the voice of Him who spake."[46] Then Christ announced, "It is done!" - meaning "all that the prophets had spoken; all that was spoken (Rev. IV, 1.)."[47] Wesley thought of prophecy as some sort of picture puzzle, consisting of a vast number of individual pieces, that by themselves seem to have little significance. However, when Christ finalizes all things and announces "it is done" - that is, a new heaven, earth, and city - the entire puzzle is completed, all pieces fit together in their appropriate places, and the new Jerusalem is the last piece to fall from the eternal heaven into place. In the entire prophetic program, Christ, as Creator and Governor, is always both the Alpha and the Omega. All things begin in him and reach their end (goal) in him, and Wesley would certainly say, "He is the entelechy that drives all things from beginning to end."

Wesley observed a sharp contrast in this sub-passage, between the new Jerusalem and the second death - the holy city as the eternal receptacle for the righteous, and the second death (or lake of fire) as the eternal receptacle for the "fearful and unbelieving."[48] The latter persons are those, he claimed, who, through lack of either courage or faith, "do not overcome." To overcome the sinful world, the flesh, the devil, temptations, pain, suffering, and even death, takes both "courage and faith." Some who do not overcome are called the "abominable" - which Wesley claimed means "Sodomites."[49] Others become whoremongers, sorcerers, and idolaters.[50] One should not be judgmental of Wesley for this view. After all, he was quoting directly from the Scriptures. Moreover, the Church of England had a Homily, number XI, entitled, *AGAINST WHOREDOM AND UNCLEANNESS*, which was read periodically in the parish churches. The early Church had its Nicolaitans, and periodically they have temporary resurgences in every age. But, Wesley believed, no matter what the century, whoever practice sodomy, whoredom, sorcery, or idolatry shall "have their part in the lake."[51]

(3) XXI, 9-14 -

"And there came one of the seven angels that had the seven phials full of the seven last plagues, and talked with me, saying, come hither, I will show thee the bride, the Lamb's wife. And he carried me away in the spirit to a great and high mountain, and showed me the holy city Jerusalem, descending out of heaven from God, Having the glory of God: her window was like the most precious stone, like a jasper stone, clear as crystal; Having a wall great and high, having twelve gates, and at the gates twelve angels, and the names written thereon, which are the names of the twelve tribes of the children of Israel; On the east three gates; and on the north three gates; and on the south three gates; and on the west three gates. And the wall of the city had twelve foundations, and upon them twelve names of the twelve apostles of the Lamb."

According to Wesley, the angel of this passage was the same angel who earlier showed the prophet John the great city called Babylon.[52] Wesley enjoyed contrasts - the city of Babylon (Rev. XVII, 1f) was termed both the "great whore" and the "mother of harlots" - the new Jerusalem was termed "the bride" and the "Lamb's wife." The first was a city of vile passions - the second is a city of purity and fidelity. The first city became the dwelling place of the Antichrist - the new Jerusalem is the eternal habitation of God in the midst of his people.

Unlike Ezekiel's vision of a rebuilt Jerusalem, dominated by a new temple (Ezekiel XL - XLVIII), the new Jerusalem of the Apocalypse shall have no temple, Wesley observed - "St. John saw no temple, and describes the city far more large, glorious, and heavenly than the prophet."[53] Its glory was God's personal glory, his effulgent light. Wesley noted John's words concerning the city - "Her window." For Wesley, such an aperture was singular, and it ran about the entire circumference of the city, allowing the glorious light to all beyond the city.[54] As for the twelve angels at the twelve gates of the city, Wesley reasoned, they are "still waiting upon the heirs of salvation."[55] His idea of "waiting" is that of "serving," by screening those who go in and out. While the city is situated on a new earth - not all men gain the holy city, and there are other redeemed creatures of a lower order than man living outside the new Jerusalem - angelic watchfulness at the gates is a necessity. The textual reference to the "twelve foundations" as the twelve apostles led Wesley to affirm once more the importance of the apostolic tradition - "Figuratively showing that the inhabitants of the city built only on that faith which the apostles once delivered to the saints."[56]

(4) XXI, 15-21 -

"And he that talked with me had a measure, a golden reed, to measure the city and the gates thereof, and the wall thereof. And the city lieth foursquare, and the length is as large as the breadth: and he measured the city with the reed, twelve thousand furlongs. The length and the breadth and the height of it are equal. And he measured the wall thereof, an hundred and forty-four reeds, the measure of a man, that is, of an angel. And the building of the wall thereof was jasper: and the city was of pure gold, like clear glass. And the foundations of the wall of the city were adorned with all manner of precious stones. The first foundation was a jasper; the second, a sapphire; the third, a chalcedony; the fourth, an emerald; the fifth, a sardonyx; the sixth, a sardius; the seventh, a chrysolite; the eighth, a beryl; the ninth, a topaz, the tenth, a chrysoprase; the eleventh, a jacinth; the twelfth, an amethyst. And the twelve gates were twelve pearls; each of the gates was of one pearl: and the street of the city was pure gold, transparent as glass."

The measuring of the new Jerusalem, by the angel with a reed (*metron* in Greek or "canon" in ancient Sumerian), fascinated Wesley. The "furlong" equals six hundred and sixty feet, so we must engage in some basic calculations. "Twelve thousand furlongs" was the angel's measurement of one side's length and another twelve thousand furlongs for its height. The city was a perfect cube, "foursquare" in length and breadth and height. Its length then, in English miles, would be one thousand and five hundred miles. Its breadth would be the same, as would its height. Its cube volume in miles would then be in excess of three billion miles. Little wonder that Wesley compared the new Jerusalem to the old Jerusalem that had a total circumference of slightly more than four miles, and Alexandria with its thirty-seven and a half mile circumference, and Nineveh with its fifty mile perimeter, and Babylon with its sixty mile spread.[57] The new Jerusalem, Wesley believed, by taking these measurements literally, will far surpass the greatest cities of history, both in size and perfection. As for the old Jerusalem, Alexandria, Nineveh, Babylon, Rome, and all the other great cities, Wesley concluded - "What inconsiderable villages were all these compared to the new Jerusalem?"[58] A city this large, he argued, would be needed to house the great number of saints from all the ages.[59] In treating the subject of measurements, Wesley equated the twelve thousand furlongs with the "one hundred and forty-four reeds." The first is man's measure and the other is the angelic measure.[60] The walls were of jasper, like the present columns of Wesley Chapel, City Road, London - a Methodist foretaste of the new Jerusalem! The city was of pure gold, its walls of jasper, with pearls for gates, and precious stones for foundations. Did Wesley take these literally too? No, he didn't -

"The gold, the pearls, the precious stones, the walls, foundations, gates, are undoubtedly figurative expressions; seeing the city itself is in glory, the inhabitants of it have spiritual bodies: yet these spiritual bodies are also real bodies, and the city is an abode distinct from its inhabitants, and proportioned to them who take up a finite and determinate space. The measures, therefore, above mentioned, are real and determinate."[61]

(5) XXI, 22-27 -

"And I saw no temple therein: for the Lord God Almighty and the Lamb are the temple of it. And the city hath no need of the sun, neither of the moon, to shine on it: for the glory of God hath enlightened it, and the Lamb is the lamp thereof. And the nations shall walk by the light thereof: and the kings of the earth bring their glory into it. And the gates of it shall not be shut by day: and there shall be no night there. And they shall bring the glory and the honor of the nations into it. But there shall in nowise enter into it anything common, or that worketh abomination, or maketh a lie: but they who are written in the Lamb's book of life."

The temple motif was important to Wesley. God and Christ are the temple. They fill the new heaven and the new earth and the new Jerusalem with their glorious presence. This new state exists in eternity, and eternity is the essential property of the divine Being. They have not come into the new heaven, earth, and city. Rather the new heaven, the new earth, and the new city have come into God's Being, as into a holy temple. In the words of Wesley, "He fills the new heaven and the new earth. He surrounds the city and sanctifies it, and all that are therein. He is 'all in all.'"[62] The reference to the "kings of the earth" is taken from Isaiah LX, 3. At long last, he thought, this prophecy is fulfilled. The word "common" (verse 27), he pointed out, means "unholy." After God has triumphed over evil, he will hardly allow it to raise an ugly head again. He will jealously guard his eternal order.

(6) XXII, 1-5 -

"And he showed me a river of the water of life, clear as crystal, proceeding out of the throne of God and of the Lamb. In the midst of the street of it, and on each side of the river, is the tree of life, bearing twelve sorts of fruits, yielding its fruit every month: and the leaves of the tree are for the healing of the nations. And there shall be no more curse: but the throne of God and of the Lamb shall be in it; and his servants shall worship him, And shall see his face; and his name shall be in their foreheads. And there shall be no night there; neither is there need of a lamp, or of the light of the sun; for the Lord God will enlighten them: and they shall reign for ever."

The all important "river of the water of life" is symbolic of the "fruitful effluence of the Holy Ghost," Wesley explained. Instead of an actual river flowing through the holy city, the Holy Spirit will be everywhere present, causing the very *zoa* of divine life to perpetually spring up within the children of God. The Spirit, as leaves ground into medicine, shall heal and keep in perfect health all who enter the gates of the holy city. The Spirit will cure the ethnic and racial madness that infects good people of every nation, and he will keep the holy community free from reinfection.

The concluding exegesis by Wesley of the New Jerusalem passage is worthy of citation -

"3. *And there shall be no more curse* - But pure life and blessing; every effect of the displeasure of God for sin being now totally removed. *But the throne of God and the Lamb shall be in it* - That is, the glorious presence and reign of God. *And his servants* - The highest honour in the universe. *Shall worship him* - The highest employment.

4. *And shall see his face* - Which was not granted to Moses. They shall have the nearest access to, and thence the highest resemblance, of Him. This is the highest expression in the language of Scripture to denote the most perfect

happiness of the heavenly state. *And his name shall be in their foreheads* - Each of them shall be openly acknowledged as God's own property, and His glorious nature most visibly shine forth in them. *And they shall reign* - But who are the subjects of these kings? The other inhabitants of the new earth. For there must needs be an everlasting difference between those who when on earth excelled in virtue, and those comparatively slothful and unprofitable servants who were just saved by fire. The kingdom of God is taken by force; but the prize is worth the labour. Whatever of high, lovely, or excellent is in all the monarchies of the earth is altogether not a grain of dust, compared to the glory of the children of God. God 'is not ashamed to be called their God, for whom He hath prepared this city.' But who shall come up into His holy place? 'They who keep His commandments.'

5. *And they shall reign for ever and ever* - What encouragement is this to the patience and faithfulness of the saints, that, whatever their sufferings are, they will work out for them 'an eternal weight of glory!' Thus ends the doctrine of this Revelation, in the everlasting happiness of all the faithful. The mysterious ways of Providence are cleared up, and all things issue in an eternal Sabbath, an everlasting state of perfect peace and happiness, reserved for all who endure to the end!"[63]

The point made by Wesley, in the comment on verse 4, should not be lost in the triumph of his eloquent ending. He carefully distinguished between the saved who inherit eternal life following the great judgment. Some are saved and granted habitation in the new Jerusalem because of their excellence of virtue in the former life - they are also granted kingship to rule. But over whom? Wesley's answer - Over those, who at the great judgment, were barely saved, who in the previous life were slothful and unprofitable, and even backsliders whose bodies were handed over to death that their souls might be saved. These do not inherit the city but only the new earth surrounding the city.

So, God, who is from everlasting to everlasting, has completed his program of love, wisdom, grace, mercy, and justice, by creating, redeeming, reconciling, restoring, transforming, and recreating; justifying, regenerating, sanctifying, glorifying, until he is all in all, and all things are perfected in Christ Jesus, his only begotten Son from eternity. And the pilgrims who set their hearts and feet on his path of perfection - climbing always heavenward, with eyes clinging to Jesus Christ as the *Vorbild*, with the entelechy of the Spirit of Christ driving them onward and upward, praying, loving, forgiving, buying up time and opportunity to serve God and man, abstaining from all evil, doing good in every circumstance, having the mind of Christ as humility, possessing peace and joy in the Spirit, always being ready to meet their God and give an account of themselves - these are the pilgrims who shall endure to the end, to stand complete in the great judgment and be sentenced to eternal life in the new Jerusalem. John Wesley desired to be counted among them.

> "As sure as thou walkest with God on earth,
> Thou shalt also reign with Him in glory!"

- From Wesley's Discourse XIII on the Sermon on the Mount

Epilogue: Wesley's View of Salvation History From the Book of Revelation

"It may be proper to subjoin here a short view
of the whole contents of this book.

In the year of the world,

3940.   Jesus Christ is born, three years before the common
           computation. In that which is vulgarly called, the thir-
           tieth year of our Lord, Jesus Christ dies; rises; ascends.

A.D.

96   The Revelation is given; the coming of our Lord is de-
           clared to the seven churches of Asia, and their angels.
           Revelation I-III,

97-98   The seven seals are opened, and under the fifth the
           chronos is declared. Revelation IV-VI.
           Seven trumpets are given to the seven angels. Revela-
           tion VII-VIII.
           Century: 2nd, 3rd, 4th, 5th, the trumpet of the 1st,
           2nd, 3rd, 4th angel. Revelation VIII.

510-589   The first woe.

589-634   The interval after the first woe. Revelation IX.

634-840   The second woe.

800   The beginning of the non-chronos: many kings. Rev-
           elation IX-X.

840-947   The interval after the second woe.

847-1521   The twelve hundred and sixty days of the woman, after
           she had brought forth the man-child. Revelation XII,
           6.

947-1836   The third woe. Revelation XII, 12.

1058-1836   The time, times, and half a time; and within that pe-
             riod, the beast, his forty-two months, his number 666.
             To Revelation XIII, 5.

1209   War with the saints: the end of the chronos. Revela-
             tion XIII, 7.

1614   An everlasting gospel promulged. Revelation XIV, 6.

1810   The end of the forty-two months of the beast; after
             which, and the pouring out of the phials, he is not, and
             Babylon reigns queen. Revelation XV-XVI.

1832   The beast ascends from the bottomless pit. Revelation
             XVII-XVIII.

1836   The end of the non-chronos, and of the many kings;
             the fulfilling of the word, and of the mystery of God;
             the repentance of the survivors in the great city; the
             'end of the little time,' and of the three times and a
             half; the destruction of the beast; the imprisonment of
             Satan. Revelation XIX-XX.

Afterward: The loosing of Satan for a small time; the beginning of
             the thousand years' reign of the saints; the end of the
             small time. Revelation XX.

             The end of the world; all things new. Revelation XX-
             XXII."[64]

WESLEY'S PRAYER

"O God, whatsoever stands or falls, stands or falls by Thy Judgment. Defend Thy own truth! Have mercy on me and my readers!
To Thee be glory for ever!"[65]

Notes

1   METHODIST HYMNAL, *op.cit.*, 1078.
2   NICENE FATHERS, *op.cit.*, II, 436.
3   WORKS, *op.cit.*, XIV, 323.
4   EXPLANATORY NOTES NEW TESTAMENT, *op.cit.*, 549.
5   WORKS, *op.cit.*, VI, 242.
6   EXPLANATORY NOTES NEW TESTAMENT, *op.cit.*, 1040.
7   *Ibid*, 1041.
8   *Loc.cit.*
9   *Ibid*, 898-899.
10   *Ibid*, 851.
11   WORKS, *op.cit.*, V, 179. Cf. JOURNAL, *op.cit.*, V, 176, 337.
12   EXPLANATORY NOTES NEW TESTAMENT, *op.cit.*, 1042.
13   *Loc.cit.*
14   WORKS, *op.cit.*, VI, 289.
15   *Ibid*, VI, 290
16   *Loc.cit.*
17   *Loc.cit.*
18   *Loc.cit.*
19   *Loc.cit.*
20   *Loc.cit.*
21   *Ibid*, VI, 291.
22   *Loc.cit.* Cf. ANTE-NICENE FATHERS, *op.cit.*, I, 566. Irenaeus, in his *AGAINST HERESIES*, made the distinction between substance and form in the new creation.
23   *Ibid*, VI, 292.
24   *Loc.cit.*
25   *Ibid*, VI, 292-293.
26   *Ibid*, VI, 293.
27   *Ibid*, VI, 294.
28   *Ibid*, VI, 295.
29   *Ibid*, VI, 245.
30   *Ibid*, VI, 246.
31   *Ibid*, VI, 247.
32   *Loc.cit.*
33   *Loc.cit.*
34   *Ibid*, VI, 248.
35   *Ibid*, VI, 248-249.
36   *Ibid*, VI, 249.
37   NICENE FATHERS, *op.cit.*, II, 436.

38 Meagher, *op.cit.*, II, 2304.
39 William Haller, *THE RISE OF THE PURITANS* (New York: Harper Torchbooks, 1957), 191.
40 EXPLANATORY NOTES NEW TESTAMENT, *op.cit.*, 1042. Cf. EXPLANATORY NOTES OLD TESTAMENT, *op.cit.*, III, 2118.
41 *Loc.cit.*
42 *Loc.cit.*
43 *Ibid*, 844, 703. In his comment of Ephesians I, 11, Wesley calls this city the "heavenly Canaan."
44 *Ibid*, 1042.
45 *Loc.cit.*
46 *Loc.cit.* and 703. Cf. WORKS, *op.cit.*, VI, 181, 296.
47 *Loc.cit.*
48 *Loc.cit.*
49 *Loc.cit.*
50 *Loc.cit.*
51 *Loc.cit.*
52 *Loc.cit.*
53 *Ibid*, 1044.
54 *Loc.cit.*
55 *Loc.cit.*
56 *Loc.cit.*
57 *Loc.cit.*
58 *Loc.cit.*
59 *Loc.cit.*
60 *Ibid*, 1045.
61 *Loc.cit.*
62 *Ibid*, 1046.
63 *Ibid*, 1047.
64 *Ibid*, 1051.
65 *Ibid*, 1052.

From Everlasting to Everlasting
Bibliography

A

Abbey, C.J. and Overton, J.H., *THE ENGLISH CHURCH IN THE EIGHTEENTH CENTURY* (London: Longmans and Green and Company, 1878)

A'Kempis, Thomas, *THE IMITATION OF CHRIST* (New York: Books, Inc.)

Anthelmi, *DISQUISITO DE SYMBOL ATHANASIUS* (Paris, 1693)

ARMINIAN MAGAZINE, THE (London: J. Paramour, 1784)

B

Baker, Frank (Ed), *THE WORKS OF JOHN WESLEY: LETTERS*, Volume XXV, (Oxford: The Clarendon Press, 1980)

Battenhouse, Roy W. (Ed), *A COMPANION TO THE STUDY OF ST. AUGUSTINE* (New York: Oxford University Press, 1955)

Baxter, Richard, *THE PRACTICAL WORKS OF RICHARD BAXTER* (London: Arthur Hall and Company, 1847)

Baxter, Richard, *THE SAINTS EVERLASTING REST* (11th Edition, Soho: J. Barfield, 1817)

Beveridge, William, *AN EXPOSITION OF THE THIRTY NINE ARTICLES OF THE CHURCH OF ENGLAND* (London: James Duncan, 1830)

Beveridge, William, *ECCLESIA ANGLICANA ECCLESIA CATHOLICA* (Oxford: University Press, 1846)

Beveridge, William, *THEOLOGICAL WORKS* (Oxford: John Bishop of St. Asaphs, 1842)

Body, Alfred H., *JOHN WESLEY AND EDUCATION* (London: The Epworth Press, 1936)

BOOK OF COMMON PRAYER, (New York: Henry Frawde, 1987)

Borgen, Ole, *JOHN WESLEY ON THE SACRAMENTS* (Nashville: The Abingdon Press, 1972)

Bourke, Vernon J., *THE POCKET AQUINAS* (New York: Washington Square Press, 1968)

Brewster, Edwin, *CREATION: A HISTORY OF NON-EVOLU-TIONARY THEORIES* (Indianapolis, Bobbs-Merrill Company, 1927)

Bull, George, *DEFENSIO FIDEI NICAENAE* (Oxford: John Henry Parker, 1851)

Bull, George, *HARMONIA APOSTOLICA* (Oxford: John Henry Parker, 1842)

Burleigh, John H.S., (Ed), *LIBRARY OF CHRISTIAN CLASSICS*, VI, (Philadelphia: Westminster Press, 1953)

Burn, A.F., *THE ATHANASIAN CREED AND ITS EARLY COMMENTARY* (Cambridge: The University Press, 1896)

Burtner and Chiles (Eds), *A COMPEND OF WESLEY'S THE-OLOGY* (New York: Abingdon Press, 1954)

C

Calkins, Mary W., *BERKELEY: ESSAYS, PRINCIPLES, DIA-LOGUES* (New York: Charles Scribner's Sons, 1957)

Calvin, John, *COMMENTARY ON A HARMONY OF THE EVANGELISTS, MATTHEW, MARK, AND LUKE* (Grand Rapids: William B. Eerdmans Publishing Company, 1957)

Calvin, John, *THE INSTITUTES OF THE CHRISTIAN RELI-GION* (Philadelphia: The Presbyterian Board of Publication, 1936)

Cameron, R.M., *THE RISE OF METHODISM: A SOURCE BOOK* (New York: The Philosophical Library, 1954)

Cannon, William R., *THE THEOLOGY OF JOHN WESLEY* (New York: Abingdon Press, 1946)

Cardwell, Edward (Ed), *THE TWO BOOKS OF COMMON PRAYER* (Oxford: The University Press, 1841)

Carter, Henry, *THE METHODIST HERITAGE* (London: The Epworth Press, 1951)

Carter, K. Codell, *A CONTEMPORARY INTRODUCTION TO LOGIC* (London: Glencoe Press, 1977)

Cell, George C., *THE REDISCOVERY OF JOHN WESLEY* (New York: Henry Holt and Company, 1935)

CERTAIN SERMONS OR HOMILIES APPOINTED TO BE READ IN CHURCHES IN THE TIME OF QUEEN ELIZABETH OF FAMOUS MEMORY (London: SPCK, 1843)

Clark, Elmer, *THE WARM HEART OF WESLEY* (New York: The Association of Methodist Historical Societies, 1950)

Clarke, Samuel, *A DEMONSTRATION OF THE BEING AND ATTRIBUTES OF GOD* (Stuttgart und Bad Cannstatt 8, Friedrich Frommann Verlag, 1964)

Clarke, Samuel, *DISCOURSE CONCERNING UNCHANGEABLE OBLIGATIONS OF NATURAL RELIGION AND THE TRUTH OF CERTAINTY OF THE CHRISTIAN REVELATION* (London: W. Botham, 1706)

Cohen, A., *EVERYMAN'S TALMUD* (New York: Schoken Books, 1975)

Cohn, Norman, *THE RISE OF THE MILLENNIUM* (Fairlawn: Essential Books, 1957)

Colie, Rosalie L., *LIGHT AND ENLIGHTENMENT* (Cambridge: The University Press, 1957)

Cox, W.L.P., *THE CHURCH OF ENGLAND AS CATHOLIC AND REFORMED* (London: Elliot Stock, no date)

Curnock, Nehemiah (Ed), *THE JOURNAL OF THE REV. JOHN WESLEY, A.M.* (London: The Epworth Press, 1960)

D

Deschner, John, *WESLEY'S CHRISTOLOGY* (Dallas: Southern Methodist University Press, 1960)

E

Eaton, Ralph M. (Ed), *DESCARTES SELECTIONS* (New York: Charles Scribner's Sons, 1955)

Edwards, Maldwyn, *FAMILY CIRCLE* (London: The Epworth Press, 1949)

Edwards, Maldwyn, *JOHN WESLEY AND THE EIGHTEENTH CENTURY* (London: The Epworth Press, 1955)

Edwards, Maldwyn, *MY DEAR SISTER* (Manchester: Penworth Ltd.)

Edwards, Paul (Ed), *ENCYCLOPEDIA OF PHILOSOPHY* (New York: The Macmillan Company, 1967)

Evans, J.M., *PARADISE LOST AND THE GENESIS TRADITION* (Oxford: The Clarendon Press, 1968)

F

Fairweather, A.M. (Ed), *THE LIBRARY OF CHRISTIAN CLASSICS*, (Philadelphia: Westminster Press, 1954)

Fairweather, Eugene R. (Ed), *THE LIBRARY OF CHRISTIAN CLASSICS*, (Philadelphia: Westminster Press, 1956)

FIRST PRAYER BOOK OF KING EDWARD VI (London: Griffith Harran Browne Company, Ltd., 1911)

Fitchett, W.H., *WESLEY AND HIS CENTURY* (New York, Cincinnati: Abingdon Press, 1922)

Foakes-Jackson, F.J., *ANGLICAN CHURCH PRINCIPLES* (New York: The Macmillan Company, 1924)

G

Gasquet, Francis A., *EDWARD VI AND THE BOOK OF COMMON PRAYER* (Third Edition; London: John Hodges, 1891)

GENEVA BIBLE, THE (London: 1615)

Gill, Frederick C., *CHARLES WESLEY THE FIRST METHODIST* (New York: Abingdon Press, 1964)

Gwatkin, H.M., *THE ARIAN CONTROVERSY* (New York: Randolph Company)

H

Haddal, Ingvar, *JOHN WESLEY* (New York: Abingdon Press, 1960)

Haller, William, *THE RISE OF PURITANISM* (New York: Harper Torchbooks, 1957)

Hardy, T.D., *THE ATHANASIAN CREED* (London: 1872)

Henson, H.H., *STUDIES IN ENGLISH RELIGION IN THE SEVENTEENTH CENTURY* (London: John Murray, 1903)

Herbermann, Charles G. (Ed), *THE CATHOLIC ENCYCLOPEDIA* (New York: Robert Appleton Company, 1907)

Higham, Florence, *CATHOLIC AND REFORMED* (London: SPCK, 1962)

Holden, H.W., *JOHN WESLEY IN COMPANY WITH HIGH CHURCHMEN* (London: The Church Press Company, 1870)

HOMILIES (London: SPCK, 1843)

HOMILIES (London: The Prayer Book and Homily Society, 1833)

Horne, T.H., *CONCISE HISTORY AND ANALYSIS OF THE ATHANASIAN CREED* (London: 1837)

Hough, L.H., *ATHANASIUS: THE HERO* (New York: Abingdon Press, 1906)

Hunter, A.M., *THE TEACHINGS OF CALVIN* (London: James Clarke and Company, Ltd., 1954)

HYMNAL OF THE METHODIST EPISCOPAL CHURCH (New York: Phillips and Hunt, 1884)

I

THE INTERPRETER'S DICTIONARY OF THE BIBLE (Nashville: The Abingdon Press)

J

Johnson, Oliver A., *ETHICS* (New York: Holt, Rinehart and Winston)

Jones, Rufus M., *THE CHURCH'S DEBT TO HERETICS* (New York: George H. Doran Company, 1924)

Jones, W.T., *THE CLASSICAL MIND* (New York: Harcourt, Brace, World, Inc., 1964)

K

Kaufmann, Walter, *PHILOSOPHIC CLASSICS* (Englewood Cliffs: Prentice Hall, 1960)

Ken, Thomas, *PROSE WORKS* (London: Griffith, Farran, and Okeden)

L

Lawson, John, *METHODISM AND CATHOLICISM* (London: SPCK, 1954)

Lee, Umphrey, *JOHN WESLEY AND MODERN RELIGION* (Nashville: Cokesbury Press, 1936)

Lindstrum, Harald, *WESLEY AND SANCTIFICATION* (London: The Epworth Press, 1946)

Locke, John, *SOME THOUGHTS CONCERNING EDUCATION* (Cambridge: The University Press, 1913)

Luccock, Halford E., *THE STORY OF METHODISM* (New York: Abingdon Press, 1949)

Luther, Martin, *TISCH REDEN* (Stuttgart: Ahrenfried Klotz, 1956)

M

Malebranche, Nicolas, *THE SEARCH AFTER TRUTH* (Columbus, Ohio: Ohio State University Press, 1980)

Marrou, Henri, *ST. AUGUSTINE AND HIS INFLUENCE THROUGH THE AGES* (New York: Harper Torchbooks, 1957)

McConnell, Francis J., *JOHN WESLEY* (New York: Abingdon Press, 1939)

McKeon, Richard, *SELECTIONS FROM MEDIEVAL PHILOSOPHERS* (New York: Charles Scribner's Sons, 1958)

McNeill, John T. (Ed), *LIBRARY OF CHRISTIAN CLASSICS* (Philadelphia: Westminster Press, 1960)

Meagher, Paul K. (Ed), *ENCYCLOPEDIC DICTIONARY OF RELIGION* (Washington, D.C.: Corpus Publications, 1979)

Monk, Robert C., *JOHN WESLEY: HIS PURITAN HERITAGE* (Nashville: Abingdon Press, 1966)

More, Paul E. and Cross, Franklin L., *ANGLICANISM* (Milwaukee: Morehouse Publishing Company, 1935)

N

Naglee, David I., *FROM FONT TO FAITH: JOHN WESLEY ON INFANT BAPTISM AND THE NURTURE OF CHILDREN* (Berne: Peter Lang Company, 1987)

Niesel, Wilhelm, *THE THEOLOGY OF CALVIN* (Philadelphia: Westminster Press, 1956)

O

Overton, J.H., *THE NONJURORS* (London: Smith, Elder and Company, 1902)

P

Portalie, Eugene, *A GUIDE TO THE THOUGHT OF SAINT AUGUSTINE* (Chicago: Henry Regnery, 1960)

Prince, John W., *JOHN WESLEY ON RELIGIOUS EDUCATION* (New York: The Methodist Book Concern, 1926)

PROCEEDINGS OF THE WESLEY HISTORICAL SOCIETY, Volume XXIX, (London)

Proctor, Francis, *THE BOOK OF COMMON PRAYER* (London: The Macmillan Company, 1955)

R

Richardson, Cyrus C. (Ed), *LIBRARY OF CHRISTIAN CLAS-SICS*, I, (Philadelphia: Westminster Press, 1953)

Robinson, John M., *AN INTRODUCTION TO EARLY GREEK PHILOSOPHY* (New York: Houghton, Mifflin Company, 1968)

Ross, W.D. (Ed), *ARISTOTLE: SELECTIONS* (New York: Charles Scribner's Sons, 1955)

S

Schaff, Philip (Ed), *THE ANTE-NICENE FATHERS* (Buffalo: The Christian Literature Company, 1887)

Schaff, Philip (Ed), *THE NICENE AND POST-NICENE FA-THERS OF THE CHRISTIAN CHURCH* (Buffalo: The Christian Literature Company, 1886)

Shedd, William, G.T., *A HISTORY OF CHRISTIAN DOCTRINE* (New York: Charles Scribner, 1864)

Shepherd, T.B., *METHODISM AND THE LITERATURE OF THE EIGHTEENTH CENTURY* (London: The Epworth Press, 1940)

Smith, T.V., *ARISTOTLE TO PLOTINUS* (Chicago: University of Chicago Press, 1956)

Smith, T.V., *FROM THALES TO PLATO* (Chicago: University of Chicago Press, 1956)

Snowden, Rita F., *SUCH A WOMAN* (London: The Epworth Press, 1963)

Southey, Robert, *THE LIFE OF WESLEY* (London: Oxford University Press, 1925)

Starkey, Lycurgus, M., *THE WORK OF THE HOLY SPIRIT* (New York: Abingdon Press, 1962)

Stevens, Abel, *THE HISTORY OF THE RELIGIOUS MOVE-MENT OF THE EIGHTEENTH CENTURY CALLED METHODISM* (New York: The Methodist Book Concern)

Stokes, Mack, *THE BIBLE IN THE WESLEYAN HERITAGE* (Nashville: Abingdon Press, 1979)

Sugden, E.H., *THE STANDARD SERMONS OF JOHN WESLEY* (London: The Epworth Press, 1956)

T

Taylor, Jeremy, *HOLY LIVING AND HOLY DYING* (New York: Appleton Company, 1847)

Telford, John (Ed), *THE LETTERS OF THE REV. JOHN WESLEY* (London: The Epworth Press, 1931)

Todd, John M., *JOHN WESLEY AND THE CATHOLIC CHURCH* (London: Hodder and Stoughton, 1958)

Tyerman, Luke, *THE LIFE AND TIMES OF THE REV. JOHN WESLEY*, M.A. (London: Hodder and Stoughton, 1876)

V

Von Loewenich, Walter, *VON AUGUSTIN ZU LUTHER* (Wittenberg: Luther-Verlag, 1959)

W

Wainwright, Arthur W., *THE TRINITY IN THE NEW TESTAMENT* (London: SPCK, 1962)

Weigle, Luther A., *THE GENESIS OCTAPLA* (New York: Thomas Nelson)

Weigle, Luther A., *THE NEW TESTAMENT OCTAPLA* (New York: Thomas Nelson)

Wesley, John, *A CHRISTIAN LIBRARY* (London: T. Cordeux, 1819)

Wesley, John, *A SURVEY OF THE WISDOM OF GOD IN THE CREATION: OR A COMPENDIUM OF NATURAL PHILOSOPHY* (Third Edition, American; New York: Bangs and Mason, 1823)

Wesley, John, *A SURVEY OF THE WISDOM OF GOD IN THE CREATION: OR A COMPENDIUM OF NATURAL PHILOSOPHY* (London: J. Fry and Company, 1777)

Wesley, John, *A COMPENDIUM OF NATURAL PHILOSOPHY: -BEING A SURVEY OF THE WISDOM OF GOD IN THE CREATION* (London: Thomas Togg and Son, 1836)

Wesley, John, *EXPLANATORY NOTES UPON THE OLD TESTAMENT* (Salem, Ohio: Schmul Publishers, 1975)

Wesley, John, *EXPLANATORY NOTES UPON THE NEW TESTAMENT* (London: The Epworth Press, 1976)

Wesley, John, *WORKS* (Grand Rapids: The Zondervan Publishing House)

Wesley, John, *THE SUNDAY SERVICE OF THE METHODISTS* (London: 1788)

Whateley, William, *A BRIDE BUSH, OR A WEDDING SERMON* (London: William Iaggard, 1617)

Whateley, William, *A CARE CLOTH OR A TREATISE OF THE CUMBERS AND TROUBLES OF MARRIAGE* (London: Kyngston, 1624)

Wheatley, Charles, *RATIONAL ILLUSTRATION OF THE BOOK OF COMMON PRAYER OF THE CHURCH OF ENGLAND* (Oxford: The Clarendon Press, 1802)

Wheeler, Henry, D. D., *HISTORY AND EXPOSITION OF THE TWENTY-FIVE ARTICLES OF RELIGION OF THE METHODIST EPISCOPAL CHURCH* (New York: Eaton and Mains, 1908)

Wiener, Philip P. (Ed), *LEIBNIZ: SELECTIONS* (New York: Charles Scribner's Sons, 1951)

Wild, John (Ed), *SPINOZA SELECTIONS* (New York: Charles Scribner's Sons, 1930)

Williams, Colin W., *JOHN WESLEY'S THEOLOGY TODAY* (New York: Abingdon Press, 1962)

Y

Yolton, John W., *JOHN LOCKE AND THE WAY OF IDEAS* (London: Oxford University Press, 1956)

Index